The Latin American Studies Book Series

The Latin American Studies Book Series promotes quality scientific research focusing on Latin American countries. The series accepts disciplinary and interdisciplinary titles related to geographical, environmental, cultural, economic, political and urban research dedicated to Latin America. The series publishes comprehensive monographs, edited volumes and textbooks refereed by a region or country expert specialized in Latin American studies.

The series aims to raise the profile of Latin American studies, showcasing important works developed focusing on the region. It is aimed at researchers, students, and everyone interested in Latin American topics.

Submit a proposal: Proposals for the series will be considered by the Series Advisory Board. A book proposal form can be obtained from the Publisher, Juliana Pitanguy (juliana.pitanguy@springer.com).

More information about this series at http://www.springer.com/series/15104

Luis Ignacio de Lasa · María Teresa Luiz

The Southernmost End of South America Through Cartography

Tierra del Fuego, the South Atlantic Ocean and Antarctica from the 16th to 19th Century

Springer

Luis Ignacio de Lasa
Instituto de Cultura Sociedad y Estado
Universidad Nacional de Tierra del Fuego
Ushuaia, Argentina

María Teresa Luiz
Instituto de Cultura Sociedad y Estado
Universidad Nacional de Tierra del Fuego
Ushuaia, Argentina

ISSN 2366-3421 ISSN 2366-343X (electronic)
The Latin American Studies Book Series
ISBN 978-3-030-65878-6 ISBN 978-3-030-65879-3 (eBook)
https://doi.org/10.1007/978-3-030-65879-3

This Springer imprint is published by the registered company Springer Nature Switzerland AG
The registered company address is: Gewerbestrasse 11, 6330 Cham, Switzerland

Acknowledgements

We would like to thank Ms. Florencia Lobo for the revision and proofreading of the Spanish original version and Ms. Elizabeth Ghitta for translating the book into English.

Contents

Chapter 1
Introduction

Abstract We present a research project that aims to facilitate the understanding of territorial processes and ponder the relationships between space and power, by studying historical cartography. We approach the integration of the southernmost tip of America into western referential systems, political territorial expansion projects and the capitalist system, focusing on space representations and territorialities in different moments of the process, from the 16th century to 19th century. We display the perspectives and concepts that guided this approach and the analysis methodology of cartographic sources.

Keywords Territorial construction · Space representations · Territorialities · Tierra Del Fuego · South Atlantic Ocean and Antarctica · 16th to 19th Century cartography

1.1 Thinking of the Territory in the Long Term

We approach the territorial construction of the southernmost tip of America from the theoretical perspectives proposed by cultural history and geography, the studies on the territory and the new political geography. We scrutinize the process from the discovery and exploration of the southern hemisphere until its integration into the capitalist system and the states' territorial design.

From a conceptualization of territory that includes social relationships, those of society with nature[1] and the spatial representations inextricably linked to these relations of power, we understand the territorial construction process as a result of material and immaterial dimensions articulation. The former is related to the political-economic sphere and the latter to a symbolic-cultural one (Haesbaert and Limonad 2007). From Fernand Braudel's point of view (1995), expressed in his studies on the Mediterranean world in the 16th century, the territory constitutes the spatial axis that organizes social actions through time. That is, a socially built space shows the past of societies, their cultural identities and the differentiated relations with the geographical environment (Llanos-Hernández 2010). Geographer Claude Raffestin (2011) also envisages territory as a socially produced space, appropriated,

L. I. de Lasa and M. T. Luiz, *The Southernmost End of South America Through Cartography*, The Latin American Studies Book Series,
https://doi.org/10.1007/978-3-030-65879-3_1

transformed and vested with meaning. That is, the result of an action performed by an actor carries out a project or program named as the syntagmatic actor by the author.

The concept of territory was traditionally associated to state spatiality, mainly—albeit not only—in international relations theory (Agnew and Oslender 2010). However, the current proposals of political geography shed light on different actors' interventions, relations of power and territorialities, considered as the different ways of organization and functioning of the territory, result of valorizations of space, intentionality and strategies implied in decision-making and management processes, as we said before, not only those implemented by the different States.

Cartography as an instrument of knowledge-power is a condition and is a result of the territorial construction processes. Maps play a key role in the production of spatial knowledge, in keeping certain political and functional control over the represented space and in the shaping of territorial images, and those representations are modeled to legitimate conceptions, beliefs, interests and interventions. This construction, in which within the studied contexts the scientific discourse and journeys literature also contribute, plays a large part in symbolic appropriation, since it conditions the ways of thinking about space, and in the medium and long term forges geographical imaginaries and territorial identities.

Bearing in mind the diverse functions of cartography—practical and instrumental, propositional, political and ideological—we delve into the modes in which it was used according to the interests and projects of different actors. As regards its use as a tool of discursive and visual appropriation of space, we identify strategies of territorial legitimation and propaganda that attempt to represent a "dominion space" with relatively accurate borders or to fix an image aimed at molding territorial awareness or belonging.

The continuity of certain modes of perception and (lack of) valorization of the space and its inhabitants by Europeans crystallizes in a representation matrix that describes an unknown and inhospitable territory on the edge of the world, singular for its "ecumenical eccentricity"[2] and the Otherness inhabiting it. The visions of an extreme geographical situation, harsh weather and wild nature of its native inhabitants, while shaping the identity of the southern tip as frontier territory, legitimate the character and scope of interventions, that is, specific types of domination.

The belief in the necessity of southern lands to balance out landmasses in the northern hemisphere conditioned the first interpretations of space and prompted European expansion plans. The hypothesis had a first empirical clue on the southern coast of the Magellan Strait and the lands situated to the south were depicted for more than a century as part of the alleged *Terra Australis*. Confirming the insular character of Tierra del Fuego did not weaken the expectations of another possible continent, and the search for new evidence in various regions of the Pacific persisted well into the 18th century, when the inexistence of inhabited or inhabitable lands beyond the Antarctic Circle was confirmed.

Cartography shows the continuity of the connection between the American southern tip and polar lands in the design of the first global strategies. As from the 18th century and during the early Antarctic exploration in the first decades of the 19th century, a southern Atlantic region, comprising Southern Patagonia, Tierra del Fuego, the Malvinas Islands, sub-Antarctic islands and the Antarctic Peninsula, is drafted. Within the context of intensification of the use of interoceanic routes and exploitation of pinnipeds and cetaceans, maps singularize a region in a geopolitical sense. That is, a space with strategic value in which several States' interests converge. Maps also grant growing importance to the Malvinas Islands as a port of call when navigating through the southern passages and as an operational base for hunting.

The analysis of the sources enables the recognition of the coexistence of different representations of space that result from distinct localizations, necessities and goals of the actors.[3] Local agents focus their attention on Pampean-Patagonian territories as the frontier of the Spanish Empire and then of the Republic. Conversely, before the eyes of the foreigners, the maritime coastlines and the South Atlantic Ocean were the almost exclusive centers of attention as transit zones on the way to the Pacific Ocean and an area of commercial interest, due to the broadening of benefits derived from exploiting marine faunal resources.

The cartographical discourse presents Patagonia as a space of formal dominion (Heredia 1998), colonial at first, and then claimed by the new political entities. Toponyms, illustrations and cartographical legends—"unknown land", "land inhabited by savages", "lands inhabited by barbarian indigenous nations", "deserted area"—reinforce the ambivalent visions that characterize the western imaginary of "there" as a construction of geographical otherness.[4] These references highlight that *terrae incognitae* "designates not only what is unknown, but also what is unexplored, ignored, alleged, supposed, badly known, plausible, expected, longed for, unbelievable, searched, beyond the horizon (the *plus ultra*), different, projected, anticipated, not conscious, foreign, external" (Lois 2018: 23).

Incorporating native spatial knowledge, Jesuit cartography proposed "other" interpretations of space. The corpus of maps, charts and journey accounts elaborated by the missionaries who operated at the southern frontiers of Chile and the Río de la Plata region during the 17th and 18th centuries offer tips on access to indigenous territoriality, the autochthonous strategies to control space and the use of resources during a process of deep transformations, derived from interethnic interactions. The use of Jesuit sources by colonial agents led to an intellectual appropriation of the interior of the continent yet unexplored by Europeans. It also enabled the design of a frontier policy that contributed to the settlement of the Hispanic-native relationships and to the consolidation of the commercial exchange that benefited colonial centers on both sides of the Andes (Luiz 2006, de Lasa and Luiz 2011).

During the last decades of the colonial period, the idea of a territory barely known and of uncertain utility justified the limited advance of Hispanic-"criolla" colonization. Likewise, the idea of "nobody's land", underlying European and North American maps of the first half of the 19th century, accompanied the development of new projects of territorial expansion. This conception is expressed by differentiating two territorial entities derived from colonial dominion in Chile and the Río de la Plata region and the depiction of Patagonia and Tierra del Fuego out of any state control, as a territory inhabited by barbarian and savage peoples and, thus, available.

Within that same context of post-independence, the cartographic representation of a southern circumpolarity, associated to the imperial ideology of maps' authors and promoters, was also part of the strategies of territorial legitimation. It also contributed to the configuration of a geopolitical imaginary that presents the South Atlantic Ocean as one of the circumpolar regions on which the British influence projected in the southern hemisphere.

Symbolic and effective integration of southern territories to the dominion of national states over the second half of the 19th century rested on an ideology that re-creates contents of the secular representation of the "inhospitable land" as a "desert", now ready to be conquered and transformed. Cartographical production took part in the two-fold process. On the one hand, a re-territorialization reinforces a national territoriality transforming indigenous territories into public lands and preparing them for an economic exploitation and, on the other hand, a de-territorialization process removes the memory of the indigenous territoriality.

Sources provide information in order to access these processes and the ways of organization and functioning of the territory, derived from specific ways in which they used and controlled space with political, social and economic purposes. Considering the actors that produce the territory with a certain goal in mind, we identify territorialities of several origins, with particular contents and logics, coexisting with different degrees of conflict. We define the specific rationality of a space of growing strategic and economic importance. Indigenous territoriality was subsumed into the political systems of the new republics, and border disputes among them were settled through diplomatic negotiations between 1881 and 1902. However, the juxtaposition of imperial and national state territorialities in the South Atlantic Ocean gave place to power rivalries and a sovereignty dispute that is still unresolved. Apart from the state territorialities, we take into account those defined by the missionaries' interventions—Anglican and Catholic—and those of businesspersons related to gold mining, sheep farming, transport, international trade and finance. Economic agents' interventions and networks organized in strong livestock and trading companies explain the territorial dynamics of the Santa Cruz region, the Magellan Strait, Tierra del Fuego and Malvinas, between 1880 and 1920 (Martinic 1992: Barbería 1991 and 1996; Harambour-Ross 2016).

1.2 Methodological Specifications

Social research traditionally limited the use of maps to "illustrate", "show" or "demonstrate" the contents of the argument (Quintero 2000). When it comes to historiography, it was mainly built over the dialogue with written documents. However, the studies on cartography history have led to appreciate its potential as "inherently rhetorical images" (Harley 2005) or eloquent registers that, when asked from pertinent categories, complete and enrich other sources' contents.

The intense pondering developed over the last decades on the theory and epistemology of cartography has contributed to renewing the approaches, something that is reflected in the works referred to the ways of organizing knowledge and the spatial experience, political and ideological assumptions and interests that influenced the production of maps in the different historical contexts.[5]

After the problematization of the assumptions of accuracy, objectivity and neutrality, the map is not still considered as "the mirror of the world, reduced through scales" (Silvestri 2011). Instead, it is viewed as a cultural product that conveys information on the society it operates in, a visual and text device that leads to a geographical data framework, contents of the imaginary, thinking systems and conceptual scaffolding from which the world is depicted. As Lois remarks (2009), recent studies "explicitly assume that the map articulates an interpretation of certain spatial relations and, though keeping certain links (clearly, not mirror-like) to an empirical reference, it is mainly the result of a social and historically defined process rather than a graphic reduction through mathematics of an abstract space".

From this approach, the scrutiny of cartographic, iconic and verbal registers involved in the discourse narrative of maps enables the access to the ways in which space was perceived, interpreted and appreciated according to certain goals. Also, to broaden the understanding of territorial formation and transformation processes, by identifying power relations expressed in spatial representations and appropriation, control, propaganda and legitimation strategies.

In this analysis, we work on maps of the world and of several regions, manuscript and printed, consulted in archives in Argentina and Spain and in digital repositories of state and private European, North American and Australian institutions. The selection of sources was performed according to the following criteria: (a) representative maps of different "types"[6]—scientific, political-propaganda and non-scientific—with enough information to identify its cultural dimension and its production, circulation and consume contexts[7]; (b) maps that enable the access to various levels of representation which, for the purpose of the present study, result independent; (c) maps corresponding to different stages of the scrutinized process, in order to visualize changes and continuities within spatial representations and territorial configurations.

The methodology built upon the dialogue between cartographical theory, studies on the history of cartography and documental sources involves two different approaches. On the one hand, an "analysis of spatial representations", in terms of relations of power that aims at uncovering what is hidden or underlying in the maps—intentionality, ideologies, motivations, strategies and technologies that support certain forms and states of power. On the other hand, an operation aimed at providing visibility and significance to what is apparent, although largely disregarded.

Upon approaching maps as "non-verbal encoded texts containing a discourse that can be verbalized from reading their constitutive signs" (Quintero 2000: 192), we identify the codes of cartographic language.[8] We do this by considering scales, projections, extreme points, latitude and longitude networks, titles, legends, toponymy, shapes, colors, drawings and pictograms. Additionally, we identify the essential elements in cartographic representations and the relationships between the physical components (relief, hydrography, natural resources) and social components (demographic, economic, political, military and cultural).

At the second stage, we organize map reading starting by the texts appearing with them, the works in which they were published or other sources that shed light on the context; that is, chronicles, general stories, journey accounts, reports and documents from state administrations. We select maps that present different geographical configurations, decisions on spatial frames and composition, selection of information, iconographies, toponyms, legends and other inscriptions. All these are conditioned by the authors' and/or promoters' goals and interests. We group maps belonging to the same period or sharing a certain feature (Harley 2005). We do this by considering the hypothesis on production conditions, the state of knowledge and the sources of spatial information, epistemological, political and ideological assumptions, maps' goals and functions, the ways in which cartographical strategies express intentionality, territorial ideologies and power rivalries. The reading itineraries intend to visualize how spatial representations, which receive feedback from practices, activate or deactivate functional or abstract territorialities.

The texts gathered in this book include the results of research projects developed at the National University of Patagonia San Juan Bosco and the National University of Tierra del Fuego, both in Argentina: "Representation of Patagonia-Tierra del Fuego space in historical cartography. The construction of the territorial identity of the southernmost tip of America within the cartographic discourse of the 16th, 17th and 18th centuries" (2010–2014) and "From 'nobody's land' to the southern border of the Republic. The territorial identity of Tierra del Fuego and the South Atlantic Ocean in the cartographical discourse. The 19th and 20th centuries" (2016–2019), respectively.

Notes

1. According to Haesbaert (2007), the territory would play the role that corresponded to the region—as an integrating concept—in classical geography. This broad concept of territory is closer to that of "geographical space" used by Milton Santos (1996), though this author privileges the economic dimension.
2. We rely on Bondel's concept (2004).
3. It is necessary to differentiate the concept of representations which, from the perspective of cultural history and geography, refers to the different forms through which communities and individuals interpret their relationships with nature and the social world, build meanings and produce reality (Chartier 1996; Bailly 1989), from the ordinary use of the concept of representations in the traditional cartographical theory that turns to the correspondence between material universe and symbolic universe (Quintero 2000).
4. Staszak (2012) uses the concept of the western imaginary of "there" in order to examine the manners in which social groups understand and tame the geographical otherness while they become aware of their identity. "There" is not far but the Otherness, which is different.
5. On the assumptions and approaches that inspire current proposals, see among others, Harley 2005, Penhos 2005, Kitchin and Perkins 2009, Lois 2009.
6. The classification is defined according to the producer/promoter (cartographic individuals or institutions), the readers/target users, the functions or aims of the map.

7. Upon context analysis, we identify producers (authors, editors and promoters), sources (selection and data confirmation), target individuals, audiences or sectors (political, commercial, scientific agents), state of geographical knowledge, philosophical and epistemological assumptions, representation regimes and cartographical practices.
8. According to Harley (2005), iconic, linguistic, numeric and spatial codes.

References

Agnew J, Oslender U (2010) Territorialidades superpuestas, soberanía en disputa: lecciones empíricas desde América Latina. Tabula Rasa, Bogotá 13:191–213

Bailly AS (1989) Lo imaginario espacial y la geografía. En defensa de la geografía de las representaciones. Anales de Geografía de la Universidad Complutense, Madrid 9:11–19

Barbería EM (1991) Chile y Argentina: una región autárquica en el sur. 1880–1920. Waxen, Publicación Científica de la Universidad Federal de la Patagonia Austral, Río Gallegos, V 4:3–43

Barbería EM (1996) Los dueños de la tierra en la Patagonia austral, 1880–1920, Universidad Nacional de la Patagonia Austral

Bondel S (2004) La Patagonia y el imaginario geográfico, ¿tan sólo una cuestión de mapas? Párrafos Geográficos, III 3:1–10

Braudel F (1995) *El Mediterráneo y el mundo mediterráneo en la época de Felipe II*. México, Fondo de Cultura Económica

Chartier R (1996) *El mundo como representación. Historia cultural: entre práctica y representación*, Buenos Aires, Gedisa

De Lasa LI, Luiz MT (2011) Representaciones del espacio patagónico. Una interpretación de las cartografías jesuíticas de los siglos XVII y XVIII. In: *Cuadernos de Historia* 35, Departamento de Ciencias Históricas Universidad de Chile, pp 7–33

Haesbaert R (2007) *El mito de la desterritorialización. Del "fin de los territorios" a la multiterritorialidad*, Río de Janeiro, Bertrand Brasil

Haesbaert R, Limonad E (2007) El territorio en tiempos de globalización. Revista Electrónica de Ciencias Sociales 1–2(4):39–52

Harambour-Ross A (2016) Sheep Sovereignties: The Colonization of the Falkland Island/Malvinas, Patagonia, and Tierra del Fuego, 1830s–1910s. Oxford University Press, Oxford Research Encyclopedia of Latin American History

Harley JB (2005) *La nueva naturaleza de los mapas*. Ensayos sobre la historia de la cartografía, México, Fondo de Cultura Económica

Heredia P (1998) "Cartografías imperiales. Notas para una interpretación de los discursos cartográficos del siglo XVIII en el Cono Sur", *Silabario*. Diseños imperiales y espacios culturales en América Latina, Revista de Estudios y Ensayos Geoculturales, I 1:77–91

Kitchin R, Perkins C (2009) Thinking about maps. In: Dodge M, Kitchin R, Perkins C (Eds) *Rethinking Maps*, London, Taylor & Francis

Llanos-Hernández L (2010) El concepto de territorio y la investigación en las ciencias sociales. Agricultura, Sociedad y Desarrollo 7(3):207–220

Lois C (2009) Imagen cartográfica e imaginarios geográficos. Los lugares y las formas de los mapas en nuestra cultura visual. *Scripta Nova. Revista Electrónica de Geografía y Ciencias Sociales*, Universidad de Barcelona, XIII (298)

Lois C (2015) El mapa, los mapas. Propuestas metodológicas para abordar la pluralidad y la inestabilidad de la imagen cartográfica. Geograficando 11, (1)

Lois C (2018) Terrae incognitae. *Modos de pensar y mapear geografías desconocidas*, Buenos Aires, Eudeba

Luiz MT (2006) *Relaciones fronterizas en Patagonia.* Ushuaia, Hanis - Universidad Nacional de la Patagonia San Juan Bosco, La convivencia hispano-indígena a fines del período colonial
Martinic M (1992) Historia de la Región Magallánica. Universidad de Magallanes, Punta Arenas
Penhos M (2005) *Ver, conocer, dominar*. Imágenes de Sudamérica a fines del siglo XVIII, Buenos Aires, Siglo XXI
Quintero S (2000) Pensar los mapas. Notas para una discusión sobre los usos de la cartografía en la investigación social. In: Escolar, C. (comp.), *Topografías de la investigación. Métodos, espacios y prácticas profesionales*, Buenos Aires, Eudeba, pp 187–217
Raffestin C (2011) *Por una geografía del poder*. México, El Colegio de Michoacán
Santos M (1996) *Metamorfosis del espacio habitado.* Oikos-Tau, Barcelona
Silvestri G (2011) *El lugar común. Una historia de las figuras de paisaje en el Río de la Plata*, Buenos Aires, Eudeba
Staszak JF (2012) La construcción del imaginario occidental del "allá" y la fabricación de las "exótica": el caso de los *toi moko* maoríes. In: Lindón A, Hiernaux D (directores), *Geografías de lo imaginario*, Barcelona, Anthropos – México, Universidad Autónoma Metropolitana, Iztapalapa, pp 177–203

Chapter 2
The First Territorialization of Southern Lands and Seas in Modern Cartography

Abstract Upon the discovery of the Magellan Strait, Renaissance authoritative sources of knowledge considered the adjacent lands as potential evidence of the existence of a fifth continent stretching to the Pole. We examine the outset of the territorial construction of America's southernmost end regarding these alleged polar lands within the debate on the geographical configuration of southern lands and the nature of said territory. We considered the role played by maps in this first territorialization and in featuring a representation framework of southern space as the border of the ecumene and the passage to other worlds. We explore formal and symbolic integration of Patagonia, Tierra del Fuego and *Terra Australis* into colonial dominion and its inclusion in the imperial designs built upon the novel sketch of the world.

Keywords Territorial construction · Tierra del Fuego · *Terra Australis* · Cartography 16th century

2.1 Introduction

In light of theoretical statements and methodological perspectives provided by studies on cartography history and recent studies on territory, in this chapter we delve into the ways of producing spatial knowledge and the first strategies of territorialization of America's southernmost end and the unknown adjacent lands.[1]

The analysis of cartographic, iconic and verbal records implied in the practices of the 16th century map-making,[2] on the basis of other documentary sources,[3] attempts to show the outset of the process of ecumenization of the southern hemisphere[4] and the territorial construction of southern lands and seas. Nonetheless, such analysis acknowledges the debate on possible geographical configurations and interpretations of the nature of space and its inhabitants. Considering the discovery of Tierra del Fuego as a landmark within the debate, we scrutinize the role played by cartography as a condition and a result of explorations and the first jurisdictional organization in South America. Assuming colonial territorialities constitute ways of territory appropriation and functioning, we inquire into the specific manners of space appraisal,

L. I. de Lasa and M. T. Luiz, *The Southernmost End of South America Through Cartography*, The Latin American Studies Book Series,
https://doi.org/10.1007/978-3-030-65879-3_2

management and significance, from the perspective of European interests, needs and expansion purposes.[5]

2.2 The Possibility of a Southern Ecumene

At the beginning of the 16th century, concurrent with the confirmation of the feasibility of a regular connection with India by circumnavigating Africa, there was an ongoing search for a way of reaching the Far East across the Atlantic Ocean. The advances toward the southwest enabled the explorers to corroborate the geographical entity of the New World as a continent or fourth part and, also, the existence of other southern lands which were interpreted, from the prevailing geographical views of the time, as a fifth part (Lois 2008b, 2010).

The chance of a vast continent in the southern hemisphere constituted one of the issues that triggered the cosmographic speculation and the imaginative thinking within the debate on the sketch of the world.[6] Likewise, the theory of climatic zones[7] gave place to the hypothesis of a specular ecumene of the known orb—and symmetrically opposite—in the temperate zone of the southern hemisphere. Upon confirmation of the habitability of the torrid zone and the likelihood of extending navigation routes through the tropics,[8] the possibility of finding new inhabited lands to the south gained credibility and displaced the medieval belief of an inaccessible antipode continent inhabited by monsters and fabulous beings.[9]

The 16th century controversy over the existence of a vast southern continent placed a strain on the theoretical, imaginative and speculative knowledge rooted in the classical tradition and the empirical knowledge nurtured by the broadening of geographical data. In other words, it conceives modern science as "a common ground where fidelity to the Ancients, a taste for the supernatural, and an already awakened awareness of that sovereign rationality in which we recognize ourselves" (Foucault 1998: 40).

The news of Amerigo Vespucci's third voyage (1501–1502) gave an unexpected twist to explorations and, above all, to the discussion over the actual nature of the islands and lands discovered in the previous decade. In a letter describing the discoveries during the expedition John II commissioned to Gonzalo Coelho in order to find a southern passage across the sea, Vespucci introduced two ideas that transformed the conceptual framework of the time. On the one hand, he affirmed that transatlantic lands were not part of Asia, as held by Columbus, but were part of a new world—a continent itself—independent from the known orb. On the other hand, he proposed the habitability of equinoctial regions and declared having seen antipodes during his navigation in the south, who he described as savage and idolatrous.[10]

Vespucci's views spread widely and cosmographer Martin Waldseemüller captured them on the globular world map included in his *Cosmographiae introductio*, published in Strasbourg in 1507.[11] The map depicts the three parts or continents of the classical ecumene as a unique landmass and the new lands discovered in the Atlantic Ocean as islands. The biggest one—South America—is named as the New

World, fourth part or *America*, after its discoverer Amerigo Vespucci. It outlines South America up to about 42° south latitude and identifies the southernmost end reached by Vespucci[12] with a Portuguese flag. Although other southern islands are included, some of which marginalized at the edges of the map as well as the southernmost end of America,[13] these would be part of Ptolemy's *Terra Incógnita*, though Waldseemüller advocated in his work, in consonance with Vespucci, for the existence of the antipodes (Fisher and von Wieser 1907; Roa-de-la-Carrera 2002).[14]

This allusion to the existence of habitable lands in the austral temperate zone provided by the authoritative cosmography knowledge of the time might be considered one of the earliest antecedents of the southern hemisphere process of ecumenization and the territorial representation of incognito austral lands (Fig. 2.1).

Francesco Roselli's world map (ca. 1508)[15]—drawn with an oval projection which enables an image of the whole earth globe—offers a full representation of the southern hemisphere in which we can differentiate several sources. The outline of the New World, with a peninsula up to 50° south latitude based on Vespucci; Africa extending up to 35° south latitude, in accordance with Portuguese explorations; the layout of the south of Asia and adjacent islands following Ptolomy; and the idea of a vast southern island—*Antarcticus*—might also stem from classical sources (Van Duzer 2008). From the territorial construction perspective, it is worth highlighting this landmass

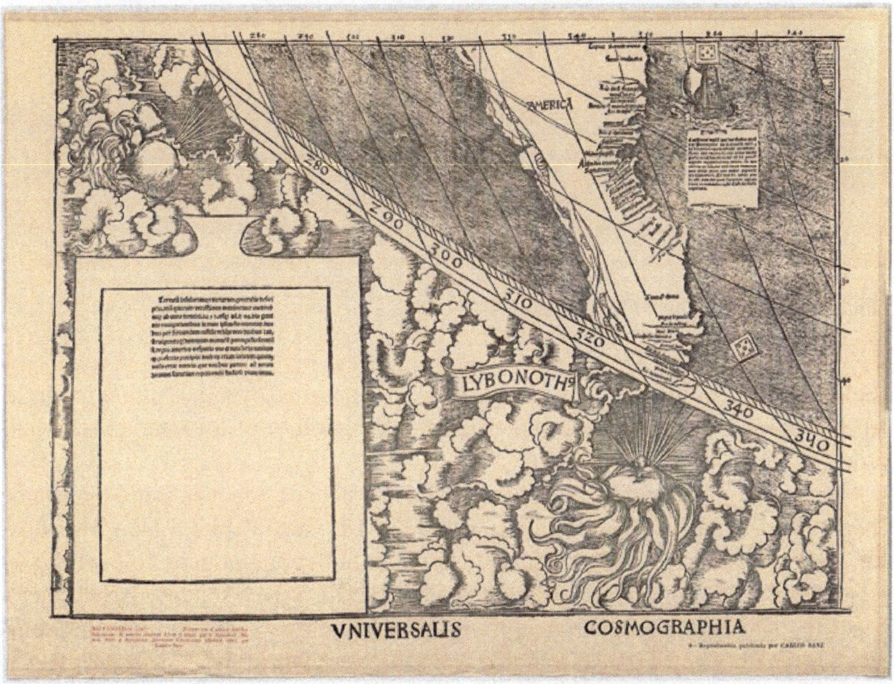

Fig. 2.1 Waldseemüller *Universalis cosmographia (1507)* detail

Fig. 2.2 Schöner *Globus* (1520)

below 50° south latitude, on whose north coast—outlined over the Antarctic Circle—there appear five map icons, suggesting the habitability of the southern temperate zone.

The globe made in 1515 by cosmographer Johannes Schöner[16] and the southern hemisphere map by Henricus Glareanus in 1513[17] also depict a New World with an insular character, which extends up to about 50° south latitude, following Waldseemüller's *Cosmographiae.* While the latter put forward a southern hemisphere of oceanic nature, Schöner sketched a ring-shaped continent stretching longitudinally within the temperate zone, separated from the New World by a strait.[18] This image of a huge continent centered in the South Pole poses the relationship between this southern landmass and the southernmost end of America, announcing a possible layout, which appears again in the 1520 map with subtle variations in toponyms (Fig. 2.2).[19]

2.3 The Discovery of Tierra Del Fuego: The Outset of the Debate on Territorialization

During the second decade of the 16th century, the Portuguese had access to the main producers of spices through the Cape of Good Hope, without abandoning the idea of finding a shorter and more direct route toward the west. In turn, the Spaniards were eagerly looking for a passage to the East through the southern portion of the New World, encouraged by the discovery of the Southern Sea by Vasco Núñez de Balboa in 1513.

From the Río de la Plata region, the first destination of the explorations promoted by the Crown,[20] Ferdinand Magellan navigated along the coast of "the mainland beyond what was already known and discovered"[21] and reconnoitered the coasts of Patagonia during the last months of 1520. The plan to get to Moluccas[22] involved "sailing southwards along the coast of the New World until they found either its termination, or otherwise the wide path through which they might reach the other Southern Sea".[23] In case there was no strait or passage due to the stretching of the known land from pole to pole,[24] "sailing along the land, they would get to the cape corresponding to the Cape of Good Hope and would discover many new lands on their way to the producers of spices, as expected".[25]

The fleet entering into the inlet discovered in October 1519 at 52° south latitude made it possible to confirm that it was the interoceanic channel. The so-called *estrecho de Once Mil Vírgenes* (Strait of Eleven Thousand Virgins) or Strait of *Patagones* opened between the mainland they were sailing along and another unknown land[26] they "assumed to be an island" and reckoned it to be inhabited because "one night they saw numerous fires on the land to the left of the strait and towards the south".[27]

The access to a new ocean—the Pacific Sea—and the first world circumnavigation provided the first empirical data to portray the whole earth globe and put an end to the debate on the diameter and extension of the *orbis terrarium.* In turn, the discovery of new lands to the south allegedly inhabited—Tierra del Fuego—posed new queries, which made it imperative to revise the premise on which the geographical conformation of the southern hemisphere and the relationship between the *ecumene* and the earth globe were imagined.

Magellan's plan design, a merge of empirical and speculative knowledge,[28] proves the coexistence of different valid conceptions of space. The search of an oceanic passage round the tip of the mainland between 35° and 50° south latitude constituted an alternative from the perspective of an insular New World and a maritime configuration of the southern hemisphere. Concurrently, the hypothesis of a strait and the existence of more lands to the Pole entailed the acceptance of one or more landmasses in the temperate and southern frigid zones.

Although the most accepted view by the end of the 15th century dismissed the existence of one or more islands comparable to the *orbis terrarum* in other hemispheres, the new discoveries in the Atlantic Ocean did not make it possible to confirm their distribution. America's depiction, independent from the known world and integrated on the orb as a "fourth part", as it appears on Waldseemüller's 1507 map of the world, paved the way to think of the existence of other lands or islands of the same nature as the southern hemisphere (O'Gorman 1995). This hypothesis was based on the classical notions of balance and symmetry, picked up by very influential medieval scholars such as Bacon and d'Ailly, and from this perspective, cartographers imagined lands that filled the oceans to balance out the northern hemisphere.

Thus, within the first two decades of the 16th century, some maps of the world or globes depicted alleged lands close to the Southern Pole. In the following decades, maps acknowledged such existence, though conferring said lands different statuses—island, archipelago or continent—and describing their territorial nature by using different cartographical resources to express the habitability of the temperate austral zone. As Portuondo remarked (2013), maps, as well as chronicles and natural histories, enabled them to transform unknown spaces into territories.

The existence of lands to the south of the Magellan Strait became the cornerstone of the hypothesis of a southern continent, and in spite of the contradictory geographical statuses conferred to these lands,[29] numerous maps of wide circulation in the mid-16th century include Tierra del Fuego in this landmass and refer to different interpretations of its territorial characteristics.

The continental configuration of southern lands is shown by means of representations that imply several shapes and dimensions, namely: an uninterrupted landmass to the Pole (Monachus 1526; Thorne 1527; Fine 1531; Schöner 1533; Mercator 1538 and 1541), a continent interrupted at about 66° south latitude or in the proximity of the frigid zone (Nancy globe 1535; Desceliers 1546; Vallard 1547; Agnese 1545; Le Testu 1555), a ring-shaped continent (Schöner 1515 and 1520; Vatican Anonymous Work 1534; Tramezzino 1554).

The uncertainty of given explanations in a period of great dynamism as regards explorations is clearly appreciated in the production of a single cartographer. Johannes Schöner portrays the latter alternative in his first 1515 and 1520 globes. He then discarded it, since his 1523 map depicts an archipelago of large islands and his 1533 globe, a whole continent with the toponym *Terra Australis.* According to Stevenson (1921), Schöner would take the lead to a new conception of the southern hemisphere. Such conception includes lands surrounding the Pole, and taking into account the similarities in configuration, extension and toponyms, he remarked that the anonymous Paris Gilt (1528), Nancy (1535) and Paris Wooden (1535) globes, as well as the Fine map of the world (1531) might be considered as part of the Schönerian School.

Oronce Fine's *Nova et Integra Universi Orbis Descriptio* (1531)[30] is the first printed map which outlines the contour of the southern land of the Magellan Strait as part of a continent which covers the South Pole (Martinic 1999) and the first one to use the toponym *Terra Australis* (Sankey 2003). The polar projection of southern and northern hemispheres enables a clear visualization of the contiguity between the southernmost tip of the New World and this huge continent. The toponym *Terra Australis* appears on this continent, followed by the reference *recenter inventa sed nondum plene cognita* (southern land recently found, but not yet fully known), and two other toponyms located in the eastern hemisphere, *Brasielie Regio* and *Regio Patalis* (Fig. 2.3).

Fig. 2.3 Fine *Nova, et Integra Universi Orbis Descriptio* (1531)

Gerardus Mercator completed the process of building up a new cartographic image of the world, which had been started by Waldseemüller and Schöner, by depicting America as a New World, separated from the Old World by a strait in the Arctic and from a southern landmass—*Quinta Pars*—by the Magellan Strait. Choosing equilibrium over symmetry,[31] Mercator argues for the importance of the distribution of the two Aristotelian elements that constituted the earth globe—land and water—and tries to balance them out in the northern and southern hemispheres (Zuber 2011). From this point of view, based on medieval and classical sources and on information on the known Tierra del Fuego and New Guinea coasts, Mercator shaped the large southern continent on his 1541 globe.[32] This configuration still appears on his 1569 map of the world.[33] This depiction raised the possibility of inferring a hypothesis on its territorial characteristics: while the Magellan region is not labeled and only three toponyms appear on the route through the strait, tropical lands to the south of the Indian Ocean are referred to as *Beach provincia aurifera* and *Maletur regnun*, taken from Marco Polo (Fig. 2.4).

Concurrently with the hypothesis that held that southern lands constituted a continent, there were other interpretations based on empirical data, which proposed an insular configuration, giving them the status of island, archipelago or "*isla de tierra*". This last category, not present when classifying lands in the 16th century, is mentioned in the "*Capitulación y asiento que se tomó con Pedro Sancho de Hoz para efectuar descubrimientos en la Mar del Sur*", of 24 January 1539.[34] If the criterion of the distinction between continent and "*isla de tierra*" was the dimension of the landmass, lack of accurate information in this regard in each case makes it impossible

Fig. 2.4 Mercator *Planisphere aux latitudes croissantes* (1569)

to determine when we are referring to one or the other one. The possibility of an error in the text transcription cannot be discounted, since the term does not appear in Sancho de Hoz's documents or in other territorial concessions. Nevertheless, we tend to explain the use of this term when referring to the particular configuration of islands of a large size, such as the one of Tierra del Fuego, as described by Alonso de Santa Cruz.

In fact, Alonso de Santa Cruz, cosmographer in the House of Trade, specifies in his Islario (c. 1545)[35] the land to the south of the strait "is proved to be an island". He criticized "Oronce and others out of Spain" and following "real and trustworthy authors", described "what was discovered of this island from fifty-three degrees to fifty-six degrees", according to the most recent information of Francisco de Camargo's expedition, commanded by Francisco de la Rivera. "The one who discovered the most of this part or island we are describing is Bishop of Plasencia's navy which, before sailing into the strait, discovered more than one hundred leagues of shores, composed of very high lands full of mountains and inhabited, as they made fires, and upon passing through the strait to the south, discovered more than sixty leagues of shores and two huge islands and an archipelago of smaller ones". Nevertheless, this latter assertion is somewhat ambiguous. Though stating it is an island—"the biggest one on earth which is the closest land to the Antarctic Pole"—he represents it on the map by continuing the eastern coast of Tierra del Fuego significantly and referring

Fig. 2.5 Santa Cruz *Mapamundi* (c. 1545)

to it as *Tierra o isla del estrecho de Magallanes* (Land or island of the Magellan Strait).[36] In addition, neither on this map nor on the 1542 map of the world—on which the northern coast of Tierra del Fuego is also longitudinally prolonged to the west, suggesting that it was a big island[37]—is the design of the coastline consistent with the information taken from Bishop of Plasencia's expedition (1539 and 1540),[38] as stated in the Islario (Fig. 2.5).

The configuration of an island of great size and close to the Southern Pole, as described by Santa Cruz, appears clearly in the production of two well-known Italian mapmakers. Battista Agnese, in his 1544 oval world maps,[39] represented an island to the south of America, located—in the different versions—between 60° and 85° south latitude. His sources would be, on the one hand, the oval world maps of F. Rosselli (1508) and S. Münster (1540 and 1545) and B. Bordone's Isolario (1528) and, on the other hand, maps derived from Padrón Real, Ribero (1529), Chaves (1533) and Cabot (1544) (Crone 1946; Wagner 1931).

Agnese, *Mapamundi Atlas* (1544), pages 28 y 29
http://bdh-rd.bne.es/viewer.vm?pid=d-1708501

In turn, cartographer Giacomo Gastaldi, Accademia Veneziana member and whose compilation skills are comparable to those of Mercator and Ortelius (Woodward 2007), represented in the "*Universale*" map of the world (1546)[40] and in the "*Carta Marina*" (1548),[41] a terrestrial mass of moderate size to the south of South America. He located it between parallel 55° south latitude and the Pole, with a limited development in longitude, mainly over the "*Oceano Meridionale*" (South Atlantic Ocean), and uses the toponym "*Tierra del Fuego Incognita*" in the northern sector[42] (Almagiá 1939). The southern island *Antarcticus* on Rosselli's map (1508), Agnese's maps of the world and some of his sources (Woodward 2007) could have inspired this construction. Reinterpretations of Gastaldi's design during 1560 decade offer different alternatives. However, P. Forlani and F. Bertelli[43] incorporated the vast southern landmass, G. Ruscelli (1561)[44] insisted on representing Tierra del Fuego as a big island (Grande 1905) (Fig. 2.6).

The controversy over the geographical nature of southern lands becomes apparent when we focus on the concurrent existence of different maps. Maps based on official Spanish cartography[45] only represent the known lands and coasts. Those which attempt to depict the southern hemisphere with the available empirical information are based on the classical premise of harmony, symmetry and balance, providing a guiding framework of design and guaranteeing accuracy. The latter set the prevailing trend, mainly during the last half of the 16th century, and express the purpose of *territorializing the austral continent* by means of different strategies. Firstly, the use of toponyms identifying places and regions[46]—names inspired on ancient and medieval sources (Mercator)—deriving from modern explorations (Schöner) or from the cartographer's fantasy (Roselli 1508 and Vatican 1534). Secondly, the inclusion of legends referring to a particular feature of the space—observed or imaginary—or

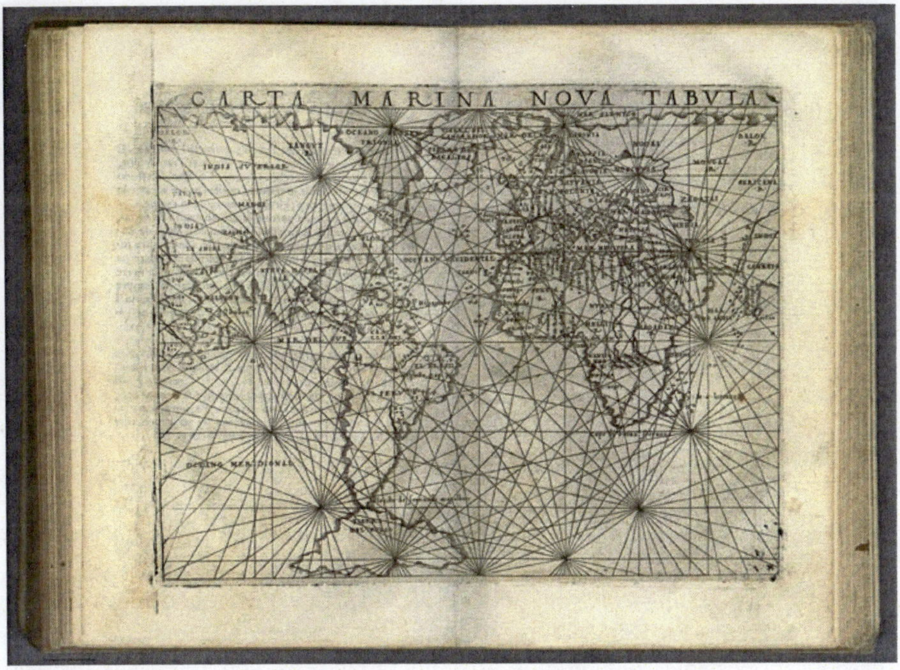

Fig. 2.6 Gastaldi *Carta Marina* (1548)

of its inhabitants, or drawing attention to riches of interest (pearls, spices, gold). Thirdly, by representing topographic details and antipode cities and by drawing the reference parallels (tropics and polar circles) indicating habitable lands between 23° 30′ and 66° 30′ south latitude. The regular use of these parallels makes it noticeable that the theory of climate zones—though somewhat discredited for the tropics due to the evidence—was still a valid criterion to order the hypothesis on the habitability of not yet explored spaces (Cattaneo 2009).

Legends and illustrations provide information on the views that interplayed in the representation of the lands—known or imagined—of the southern hemisphere. While the news of the first voyages to the Pacific Ocean[47] strengthened the belief that there was a southern continent full of riches of commercial interest, the information on the southern end of the New World and adjacent lands was less encouraging. Diego de Ribero's *Carta Universal* (1529) manuscript map of the world, inspired by the *Padrón Real*, bears elements on which to build the (de)valuation of the southernmost end of America. Describing *Tierra de Patagones* according to testimonies of members of Magellan's expedition, mainly Pigafetta's voyage account, the land is described as "barren and unprofitable" and its inhabitants as "giants covered with furs and living in the open air (Fig. 2.7)."[48]

The depiction of *Terra Australis* on the maps of the world sketched by Dieppe cartographers as of the 1540 s evinces an apparent contrast between the Magellan region and the southern lands of the Moluccas, whereas Tierra del Fuego is represented as an empty space—unknown and worthless—or a wild land at the edge of

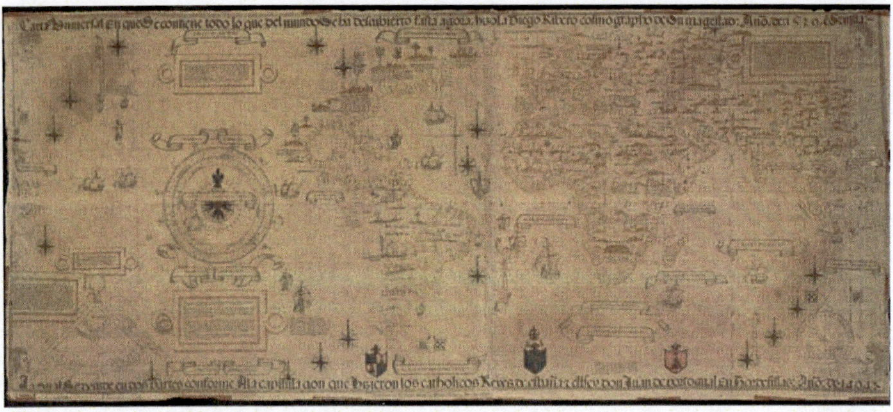

Fig. 2.7 Ribero *Carta universal* (1529) detail

the ecumene, Java la Grande, south of Sumatra, is shown as a rich and civilized territory.[49]

The *Cosmographie Universelle* atlas (1555),[50] written by French pilot and cartographer Guillaume La Testu after the expedition to Brazil sent by Francisco I in 1551, describes a possible configuration of southern lands as well as a landscape and a history. Its title presents it as a cosmography, though its contents makes it part of geographical science. There, as Lois (2008a: 78–79) very well remarks "description becomes the core of the text and grows apart from astronomical elements, getting close to history, which is not conceived as regarding only the past but as the updating of knowledge on the geography of the world".

In this work, Le Testu directed his gaze to the New World, highlighting the importance of the southern continent, which is portrayed in 12 of the 50 charts of the region as the link between the two central parts of the atlas, Asia and America. The southern landmass is depicted on empirical data available—islands found by the Portuguese and Spaniards—though the author warns about these imaginary descriptions and illustrations (Lestringant 2015). The maps depict *Terra Australis* as the continuity of the South American continent, to which they assign the same territorial nature by representing Patagonian giants to the south of the Magellan Strait. Likewise, they warn about the risks of sailing in southern latitudes of the Ocean Sea and the Southern Sea, by including figures of marine fauna of great size and shipwrecks. Le Testu proposed an imaginary circumnavigation between 30° and 50° south latitude and portrayed a territory built on expectations and the experience of the New World, of his own and of other navigators. He started his itinerary at Java la Grande and, after going down the *Mar de la India* (Indian Sea) to the west, ends at Tierra del Fuego, which he represented as an exotic, wild and violent world, in contrast to the civilized world of the southern coasts of Asia and Africa.

The depiction of imaginary territories initially aimed at explaining the morphology of the planet and, as Lestringant (2015) remarked, at filling in the gaps in geographical knowledge. However, later it intended to promote commercial interests to fund new

Fig. 2.8 Le Testu *Cosmographie Universelle* (1555)

expeditions, by locating rich civilizations in the temperate zone (Portuondo 2013). The line full of toponyms on the known coasts bolsters the territorial construction of an available territory (Fig. 2.8).

Belgian cartographer Cornelius Wytfliet, in his atlas of the New World, published in 1597, also identified distinct areas within the southern land. “To the West facing Sumatra, Java and Moluccas, to the East there is Africa and the Ethiopian Ocean, the North being at the narrowest section of the Magellan Strait and in the Patagones lands”, as depicted in *Chica sive Patagonica et Australis Terra.*[51] By using two different scales, it provides a detailed image of the Magellan region, presents various toponyms on the unique known coastline of the southern continent—Tierra del Fuego,

Fig. 2.9 Wytfliet *Chica sive Patagonica et Australis Terra* (c. 1597)

and indicates the establishment of the Rey Felipe Spanish settlement on the north coast and the coast of San Julián in the Atlantic Ocean, the arrival port most frequented before entering into the strait. The text accompanying the map still clings to the European expectations as regards the traditional Indies—the legendary Ophir, Tharsis and the Solomon Islands—and the limited knowledge of "*Chica*", the name already utilized by Mercator on his 1569 map of the world. In spite of the fact that we do not have a monographic study to clarify some morphological or semantical issues around this term, considering the poor descriptions and the fact that Patagonia is not represented to its full extent neither on this map nor in the regional maps of del Plata and Chile, we wonder whether there was value in the choice of such a name, since the term "*chico*" in Spanish, related to the semantic shift from the Latin term "*ciccum*", means "small", "little", "worthless" or "trifle" (Fig. 2.9).[52]

2.4 The Formal and Symbolic Incorporation to Colonial Dominion

The Magellan Strait offered a new access to the spice market and an alternative communication route between the peninsula and the Castilian possessions in the Pacific Ocean, purportedly less complicated than the one used by the Caribbean across the isthmus. The main goal of the expeditions following the one of Magellan-Elcano (1519–1522)[53] was to assess the navigability conditions of the route, since sailing across a strait would pose higher risks than open water sailing.

Concurrently, the southernmost end was incorporated under colonial dominion in the first jurisdictional macro-organization of the South American space derived from the extension of Francisco Pizarro governorate up to 14° south latitude and the creation of three new governorates to the south, according to the capitulations signed in May 1534 (Nocetti and Mir 1997). The governorate granted to Simón de Alcazaba comprised 200 leagues between parallels 36° and 48°, with coastlines to both oceans, which amounts to the first territorial concession on Patagonia, though actual occupation of Port Los Leones (45° south latitude) in March 1535 was abandoned due to conflicts arising among crew members. Another capitulation entered into in 1536 with the Bishop of Plasencia—Gutierre de Vargas Carvajal—on behalf of his brother Francisco de Camargo, formalized a new jurisdiction on the southernmost end. The aim was to consolidate the conquest and to inhabit the coasts of the Southern Sea from the termination of the 200 leagues granted to Pedro de Mendoza up to the Magellan Strait. Francisco de la Rivera's expedition, after Camargo's resignation, could not carry out the occupation either.[54]

On 24 January 1539, another capitulation was signed with Sancho de Hoz, in order to discover along the coast of the South Sea "the said Magellan Strait and the land stretching on its other side". In February of that same year, he was granted the title of Captain General and that of Governor with jurisdiction from the southern border of the one granted to Camargo up to the lands to the south of the strait. Although the undertaking failed due to his non-compliance with the responsibilities assumed in said document,[55] these provisions could be construed as the first legal precedent of the territorialization of Tierra del Fuego and the alleged southern continent. The semantic ambiguity of the texts proves the lack of knowledge as regards the geographic status, as the terms *land, islands, island or "isla de tierra"* are used interchangeably.[56]

While the governorates created in 1534 and 1539 were possibly derived from the maps produced in 1529 by the House of Trade cosmographer Diego de Ribero, Alonso de Santa Cruz 1542 *Carta Universal* confirmed the jurisdictional design by proposing an image of South American colonial space at the beginning of the territorial construction (Nocetti and Mir 1997).

This map is labeled with the names of the four provinces formed upon the first territorial concessions, concentrating the highest number of toponyms on the coast of the Atlantic Ocean and representing relief, hydrography, flora and fauna and some buildings in the continental interior. Icons enable the distinction of several spaces, and the lack of illustrations in the *Provincia del Estrecho* (Strait Province) could denote ignorance, lack of effective occupation, or else, less expectations as regards its significance for Spain. Magellan lands that had been given to Camargo in 1536 were now included in the Governorate of Chile, when according to the Royal Charter issued on 29 September 1554, the jurisdiction granted to Valdivia in 1548 (27°–41° south latitude) was broadened up to the Magellan Strait. His successor, Jerónimo de Alderete, received under Royal Charters signed on 29 March 1555, the same

Fig. 2.10 Gutiérrez *Americae sive qvartae orbis partis* (1562)

territorial jurisdiction and the authority to discover, conquer and inhabit the lands on the other side of the strait (Martinic B. 1981).

Official cartography reinforces these territorial concessions representing unexplored lands within the Castilian sphere of influence, that is to say, producing a territory to act in accordance to certain interests. Diego Gutiérrez, cartographer for the House of Trade, produced the *Carta de América* (Map of America) in 1562,[57] evidencing the geopolitical intentionality when including *Tierra de Patagones, Tierra de Magallanes* (Tierra del Fuego) and *Regio Patalis* (on the Pacific Ocean) as expansion spaces. The scene of a naval battle between two fleets in the South Atlantic Ocean could imply Felipe II's decision to defend this territorial demand (Kagan 2005).

Together with the reference *Gigantum regio*, the representation of a fight between two giant Patagones and a European identifies the conquest territory, relating the ideas on the natural and savage space to the humanity standard of its inhabitants, cartographically re-creating Pigafetta's description (Fig. 2.10).

Notwithstanding initial expectations, the strait route turned out to be risky, mainly due to the strong and unstable winds and the conformation of the interoceanic channel, an actual labyrinth of islands and channels, as was demonstrated by the reconnaissance trips commissioned by Chilean Governors from the Pacific Ocean. Francisco de Ulloa led the first of these trips between 1553 and 1554 and Juan Fernández Ladrillero led the second one in 1558, which turned to be more exhaustive.

Juan López de Velasco, first Cosmographer and Chief Chronicler of the Council of the Indies, elaborates his description of the Magellan Strait in Ladrillero's report and states the official judgment on the nature of the route:

Sailing across the strait, along the South Sea or the North Sea is not feasible without facing great risks, unless it is mid-summer, in the months of December and January, since there are big and relentless storms during the rest of the year, with wind blowing from the southwest and northwest, lasting ten or twelve days, and a lot of snow and constant cold weather with hail storms and threatening weather, as there are only six hours of light, and even during the Summer, those who attempted to sail across the strait had to face perils, and therefore, sailing along these coasts had always been difficult and risky because of its height; because even with the existence of plenty of good ports within the strait itself and close to it, seamen can barely make use of them due to gales and tempests, thus, those who dared to cross the strait had to bear dangers and threats.[58]

The possibility of regular communication between New Spain (Acapulco) and Philippines (Manila), upon completion of Legazpi–Urdaneta's voyage (1565), offered one more reason to avoid such risky navigation in the south. Thus, the Crown's interest focused on discovering islands and lands in the South Pacific Ocean (Rodamilans Ramos 2010).

Francis Drake's incursion into the Pacific Ocean through the Magellan Strait and his plundering Valparaíso and El Callao's ports forced Spain to shift its attention back to southern borders, on which none of the involved governorates—Chile, Tucumán and Río de la Plata—had yet actually occupied. The development of a defensive plan imposed new jurisdictional acts. Pedro Sarmiento de Gamboa was appointed Governor and Captain General of the Magellan population in 1581, and the settlements founded in 1584 on the northern border—Nombre de Jesús (In the Name of Jesus), near Cape Vírgenes, and Rey Don Felipe, in Puna Santa Ana—changed temporarily the nominal character of the Castilian jurisdiction on the southern region (Nocetti and Mir 1997).

These first territorial concessions, reinforced by the official cartography which identifies the territory with names of geographical and administrative character, and the exploratory labor, accompanied with acts of possession and giving names to places, constituted ways of intellectual and political appropriation of space. This resulted well strategically for Spain who was looking to seize control of the navigation in the south and the defense of the American coasts in the Pacific Ocean.

2.5 Imperial Designs on the New World Image

From the perspective of geometric aesthetic, Renaissance cartographers conceived an image of the orb in which a vast mass of southern land was necessary to balance out the lands in the northern hemisphere and guarantee symmetry of shapes. This conception gained high credibility with the work of Abraham Ortelius, whose map of the world held sway on the representation of the world for half a century. The map of the world *Typus Orbis Terrarum*, published in the atlas *Theatrum Orbis Terrarum* (Amberes 1570), had nine editions up to 1584 and other 16 with slight changes between 1587 and 1624, and the map entitled *America Sive Novi Orbis Nova Descriptio* had 25 editions between 1570 and 1624 (Martinic 1999). The key to this success was grounded not only in its value as regards the formal and scientific aspects of map-making but also in the authenticity added to the representation of the unknown. The broadening of the habitable world beyond the limits set by tradition enabled the construction of new spatial imaginary, merging the known and the hypothetical with the same level of certainty (Sánchez 2010; Besse 2010) (Fig. 2.11).

Based on Mercator's tripartite conception of the world—Old World or classical ecumene, New World and southern continent—Ortelius divided the orb into three parts of mainland:

Fig. 2.11 Ortelius *Typus Orbis Terrarum* (1570)

> …the first one being the one divided by the Ancients into three parts (Africa, Europe and Asia), where mankind originated, as stated in the Holy Scripture; and the second one now called America or India Occidental (Western Indies); the third one being the austral land, which is sometimes referred to as Magellan, since very few coasts had yet been discovered[59].

This third part or the fifth continent, still unknown but plausible, was seen by the various States excluded from the Lusitanian-Hispanic distribution of the New World as possible expansion areas. This took place at a time in which the trading dynamism allowed increasing benefits from overseas undertakings, and Ortelius map of the world became the knowledge-power tool of design of imperial strategies during the last quarter of the 16th century.

France, an early advocate of the idea of freedom of the seas, contemplated the unoccupied or unexplored land in the southern hemisphere as a target for its expansive plans, though with little success. The eastern adventure of the Angos from Dieppe ended in the voyage of Jean Parmentier and his brother Raoul (1529–1530), who reached Sumatra but not any further, since the French preferred to participate indirectly in the trade of spices, positioning itself in Seville, Lisbon and Antwerp and in the European redistribution routes. Though the focus on the Atlantic Ocean was worthwhile due to the significant profits obtained from Brazil wood harvests since the 1530 s, the colony founded in Guanabara Bay—France Antarctique—would barely survive for a decade, as it was later eclipsed by Portuguese monopoly (Chaunú 1984; Sankey 2002). Even though the State did not seem to be willing to compromise resources in uncertain enterprises, some still advocated for organizing expeditions and inhabiting the alleged southern lands. Since the 1540 s, Dieppe cartographers had represented a vast southern continent ready to be conquered. But it was the work *Les Trois Mondes* (1582), by Henri Lancelot de Voisin de la Popelinière, historian and author of voyage books that directly promoted its discovery and colonization by France.[60] Although that "third world did not amount to a geographical issue and, let alone, a significant and relevant geographical entity" (Lois 2008a), the proposal elaborated in the context of the Crown political and religious crisis appears to be a particular combination of geographical knowledge, imperial aspirations and utopia. Supporting his argument on the depiction of a vast southern landmass on Mercator and Ortelius' world maps, the author comes up with the possibility of an enterprise that would enable France to acquire territorial possessions to compensate for losses in Brazil and Florida. He also imagines *Terre Australe* as a possible destination for the French who had been driven out, as a place of freedom on the edge of the world (Sankey 2003).

Meanwhile, England—who had remained apart from Spain and Portugal's double monopoly—developed as of the 70 s an expansive policy which started with war and piracy on Iberian oceanic spaces, first in the Atlantic Ocean and later venturing into the Pacific Ocean (Chaunú 1984). Although the purposes of Francis Drake's voyage (1577–1580) were not documented, the itinerary designed between August and October 1578 makes us believe that they considered the idea of sailing along *Terra Australis* from Tierra del Fuego to New Guinea and Moluccas, according to Ortelius map.[61] After having crossed the Strait in 16 days,[62] the fleet was driven by a series of storms to the southwest at 56° or 57° south latitude and upon returning to the western mouth of the Strait, one of the vessels got lost and another one went back to England. Drake's ship sailed back toward the southeast along the coasts of Tierra del Fuego until reaching Elizabeth Island, off the tip of South America, proving the non-existence of mainland neither to the south nor to the east (Andrews 1968; Spate 2004). The discovery was recorded on a 1580 map sketched by Drake himself, reproduced in several probably unauthorized manuscript copies, and on a map left by Francis Fletcher, the expedition's chaplain, on which the South American continent ends on an archipelago and includes the legend *Terra Australis bene cognita.* These maps remained under a strict secrecy policy until the end of said decade (Wallis 1984).

Drake's irruption in the Pacific Ocean demanded from the Spanish colonial authorities a greater control over the alleged only access route to the Pacific Ocean, and in 1579 Francisco de Toledo, Viceroy of Peru, commissioned an expedition to the Strait on a new reconnoitering voyage to spot possible English settlements. Pedro Sarmiento de Gamboa, who commanded this expedition, recognized the interoceanic passage and sailed toward Spain to participate in the design of a defensive plan. In spite of controversies and uncertainties as regards the nature of Tierra del Fuego, the Council of the Indies approved in 1581 an ambitious plan to establish two forts in the eastern mouth of the Strait. The expedition, led by Diego Flores de Valdés commanding a fleet of 23 vessels and a crew of 3000 people, aimed at strengthening the defense of the ports in the Pacific Ocean, and starting the colonization and fortification of the Strait. To this end, 350 colonists and 400 soldiers were assigned to Magellan forts (Luiz and Schillat 1997). The plan's cartographical grounds were found on 1578 Joan Martines' map, which adds—following Ortelius representation of the southern landmass—two elements that provide significant visibility to Tierra del Fuego: coloring and shading of coasts up to the Antarctic circle and the label of the toponym *Terra del Fuego.*[63] The use of different colors not only "differentiates what is already known for sure from what is still uncertain" (Lois 2008a) but could also be interpreted from a geopolitical point of view as involving the territorialization purpose of the region on which Hispanic dominion would become effective (Fig. 2.12).

The Spanish shift from the colonies at the Strait to the Pacific Ocean cannot be dissociated from the existence of a new maritime passage between Tierra del Fuego and the so longed for southern land. Cartographer Joan Martines himself depicts it on another map of South America, part of his 1591 *Atlas Marítimo*, as a broad channel or inlet, next to which there appears a legend that assigns its discovery to Drake's Portuguese pilot Nuño da Silva (Martinic 1998).[64]

The imperial strategy was then to resume from the Lima explorations led by Alvaro de Mendaña between 1567 and 1569, assuming that the Solomon Islands—discovered by the aforementioned—were close to New Guinea, which was considered mainland. Jesuit José da Costa, residing in Peru between 1572 and 1585, and member of the group of intellectuals serving Viceroy Toledo, refers to this issue in his *Historia natural y moral de las Indias* (1590). He states that "there are several hypotheses which affirm that there exist rich and prosperous lands in the temperate zone near the Antarctic Pole, which are not yet discovered". The aim was to keep on

Fig. 2.12 Martines *Sudamérica* (1578)

searching south for lands in an area awakening now more expectations, since Ophir and Tharsis and Java la Grande legendary references—from Dieppe School—and *Beach, Lucach and Maletur* regions on Ortelius map of the world.

Meanwhile, maps of wide distribution by the end of the 16th century and the beginning of the 17th century account for persisting debate on the existence and nature of lands in the temperate southern zone. Furthermore, they explicitly state the strategical significance of the southernmost tip of America in trading with the East since, as Padrón (2009) pointed out, economic historians remind us of the importance of Asia to the emerging global economy of early modernity, whose driving force was not Europe but China, which resulted in the great significance of Spain's trans-Pacific Ocean trade through Mexico, Lima and Magellan.

Abraham Ortelius' *Maris Pacifici* is the first printed map portraying the Pacific Ocean and one of the main regional maps by the end of the 16th century, considered the contribution of "Spanish imperial cartography", that is, part of the cartographic corpus produced by the State in order to promote its territorial and political ambitions (Padrón 2009). Produced in 1585 by the, at that time, geographer of King Philip II and published in the 1589 edition of *Theatrum Orbis Terrarum*,[65] the map shows the Crown of Castile's view of the overseas lands: the Pacific Ocean as a dominion space and New Guinea, the Philippines and the Solomon Islands, China, Japan and other possible southern lands as areas of imperial interest. It depicts the Magellan Strait as the only entrance to the Southern Sea, establishing Spanish sovereignty on this route by locating one of the settlements founded on the southern coast in 1584. It also reminds us—with the illustration of Victoria vessel—not only of the discovery of

Fig. 2.13 Ortelius *Maris Pacifici* (1589)

the biggest ocean in the world but of the validity of the target of Spanish expansion, trade with the East (Fig. 2.13).

The depiction of the Pacific Ocean on Jodocus Hondius *Americae Novissima Descriptio* (1589)[66] keeps similarities with Ortelius, though its construction as an expansion space evidences the Netherlands commercial and strategic interests. It does not include neither China nor Hispanic dominion over the Philippines and highlights places which would be targets in future Dutch expeditions: New Guinea, the Solomon Islands and island arcs presumably adjacent to the southern continent coast, in accordance with pilot Hernando Gallego's discoveries in 1570. It portrays two accesses to the Pacific Ocean, one of them within Spain's domain, the King Philip settlement at *Fretum Magallanicum,* and the other one discovered by the English to the south of Tierra del Fuego, depicted as an archipelago of three big islands.

Hondius' production also shows the concealment strategy and/or the information distortion of space information in the different configurations proposed for the southernmost end of America. Whereas during his stay in England Hondius works with Drake's information, upon his return to the Netherlands in 1593, he resumed his depiction of Tierra del Fuego as part of *Terra Australis.* As Wallis (1984) stated, his maps and globes contributed to confirm Holland's belief that Le Maire and Schouten

Fig. 2.14 Hondius *Americae Novissima* (1589)

had been the discoverers of the passage round the southern tip of South America (Fig. 2.14).

Toward the end of the 16th century, the British and the Dutch increased their presence in the south. While the British ventured in a series of privateers voyages through the Strait with more costs than profits, the Dutch reconnoitered the interoceanic channel fueled by the interest of commercial companies distributing spices to use this route to reach production centers in the East Indies (Chaunú 1984; Martinic 1999). The outcome of the first Dutch expeditions that sailed along the Magellan region in 1599, captained by Jacob Mahu, Simón de Cordes and Sebald de Weert and Oliver van Noort, demonstrated that this route—though shorter than the African one—turned out to be extremely risky. Even though voyages were suspended for more than a decade, the necessity to find an alternative navigable route to the one offered by the Strait and the Cape of Good Hope reserved by the Dutch States General to the Dutch East India Company, Hoorn Company was encouraged to move forwards. In 1616, the expedition led by William Schouten and Jacob Le Maire found a second interoceanic passage at 55° 58′ south latitude.

The corroboration of the insular nature of Tierra del Fuego upon the discovery of Cape Horn route through Drake Passage and, three years later, with the first circumnavigation of the archipelago by Captains Bartolomé and Gonzalo Nodal (1618–1619), put an end to the century-old debate on the nature of Magellan lands without weakening the hypothesis of the existence of a continental southern landmass.

2.6 Conclusions

The exploration of the southern coasts of America in order to search for a maritime passage to the East led to the elucidation of the geographical entity of the New World and, concurrently, posed new queries over the geographical nature of the southern hemisphere and the existence of a southern ecumene. The scientific debate over the depiction of the world and the habitability of the different regions was inextricably linked to the purpose and necessity of overseas expansion. Cartography produced during the 16th century evinces this closeness between knowledge and power, by expressing cosmographic and geographical conceptions from which spatial information was construed as well as political and ideological assumptions, which determined space representation.

The Magellan-Elcano expedition confirmed that the passage to the lands rich in spices that they had been looking for was a strait between known mainland (America) and other southern unknown lands (Tierra del Fuego). Lands which, from the conception of a southern hemisphere, were depicted as part of a continental landmass foreshadowed in classical and medieval sources surmised to fulfill the cosmographical necessity of equilibrium and symmetry between the two hemispheres. As of the third decade of the 16th century, maps portrayed a southern continent with different shapes and dimensions. Instead, all of them hypothesized the outlines taking into account the known coasts of Tierra del Fuego, Australia and New Guinea.

The explanations of the territorial nature of the "fifth part" or "third world" on Mercator and Ortelius maps of the world were based on theoretical speculations, creative imagination and empirical data produced by the first expeditions to the southern seas. Whereas the news of the lands that had been discovered in the South Pacific Ocean strengthened the belief that there existed a continent full of resources of commercial interest, information from the southernmost end of America appeared from the outset much less stimulating and gave a place to images of foul weather and savage inhabitants. Cartographic discourse played an important role in these representations by providing a set of references on which to build the identity of Patagonia and Tierra del Fuego, from the point of view of Europeans' beliefs and expectations, as a territory of uncertain value.

The perception of difficult habitability conditions together with dangerous navigation conditions of the Magellan route explain the slow progress of exploration and colonization. In spite of the fact that colonial dominion remained at formal level, except during the brief Spanish occupation of the northern coast of the Strait, cartography bolstered the process of territorialization of the southernmost end of the continent and adjacent lands. It inscribed in the abstract space, places, itineraries, individuals, products, landscapes and events, and thus expressing by means of several decisions as regards shape and content, the appropriation of territory and legitimating political intentions.

Confirming the insular nature of lands located to the South of the Magellan Strait weakened Spain's expectations toward the possibility of controlling interoceanic navigation and triggered the search of new more promising lands in terms of economic profitability when compared to the stark and barely inhabited Patagonia. The discovery of the route round Cape Horn, as an alternative maritime route to the one through the Strait, meant no change in the (de)valuation of the region, and severing Tierra del Fuego from an alleged southern continent deviated exploration toward other Pacific Ocean regions. Although the outcome of explorations were discouraging, the search for evidence of "another possible world" never stopped until well into the 18th century when, after verifying that lands beyond the Antarctic Circle were inhabitable, Tierra del Fuego and southern archipelagos rendered in effect the modern border of the ecumene, edge or end of the inhabited and habitable world. The idea of "end of the world" would result in the transformation of a spatial category with geographical grounds—frontier—into a metaphoric construction with strong roots in the imaginary, an idea that eventually becomes a belief and, maybe, a myth, related, as Giucci (2014) suggests, to the craving for the "*lugar límite*" or "*ultima tuhle*".

Notes

1. Territorialization processes are thought as the outcome of social relationships. Those of society with nature and the several ways through which communities interpret these relationships, with a theoretical approach which merges political, economic and cultural dimensions and takes into consideration the practices, representations and strategies used by certain groups to keep certain functional and symbolic dominion over an area of geographical space (Raffestin 1993; Haesbaert and Limonad 2007).
2. We worked on pilot charts and political, scientific and general interest maps, manuscript or printed, which could be organized within a corpus, identifying authors and sponsors, sources of information and philosophical, epistemological and political assumptions. The deployed cartographic sources are available with virtual repositories of state and private institutions in Europe, North America and Australia and, in some cases, replicated in historical cartography studies.
3. We examined chronicles, general histories, atlas, cosmographies, voyage chronicles, reports and other documents of the colonial administration.
4. We refer to new knowledge and space imaginaries construction process, switching from the classical conception of the ecumene as the known and inhabited world—Europe, Asia and the north of Africa—to the representation of the globe with different parts, continents or worlds, which gives a place to inhabited or habitable areas still unknown (Lois 2008a).
5. Mignolo (1995) and Padrón (2002) refer to the role of the representation of the abstract, geometrical and homogeneous space, adopted as natural by the Renaissance cartography, in the territorialization of the non-European world and, in turn, in the deterritorialization of Amerindian peoples.
6. From Crates of Mallus' perspective of the four parts or continents (2nd century BC), Macrobius (5th century AD) speculated on the existence of a vast southern land, independent from Africa. His work was strongly influential during the Middle Ages and a major source in cosmographic and geographical debates in the 16th century (Hiatt 2007).

7. According to this theory, proposed by Parmenides (5th century BC) and developed by Eratosthenes (2nd century BC), the globe was divided into five climatic zones: two cold ones close to the poles, two temperate intermediate ones and a hot or torrid one at both sides of the Equator up to the tropics, following the parallels of astronomy reference. The first ones considered uninhabitable due to extreme cold and excessive heat, respectively. These ideas, presented by Cardinal Pierre d'Ailly in his *Imago Mundi* (1410), became an important reference source for Renaissance intellectuals.
8. Roger Bacon (1268) and John Mandeville (1356) put forward habitability of the torrid zone (Lois 2008a).
9. The medieval imaginary and particularly the Christian cosmology which denied the existence of human beings not descending from Adam and beyond the Gospel message gave place to believing on an antipode continent inhabited by Plinio's monstrous races (century I AD) and other fantastic creatures that nurtured a great number of myths and legends (Vignolo 2003).
10. We refer to the letter sent from Lisbon to Lorenzo di Pierfrancesco de' Medici in 1502, published in Latin a year later in a pamphlet called *Mundus Novus* (Paris, 1503) and with several re-editions: 12 in 1504, 22 in the next three years and about 50 before 1550 (Levillier 1956, Romano and Tenente 1979).
11. *Universalis cosmographia secundum Ptholomaie traditionem et Americi Vespucii alioru[m]que lustratione,* St. Dié, 1507. Repository: Ibero-Amerikanischen Institus Digitalen Sammlungen.
12. Vespucci's voyages triggered a controversy over the southern latitude reached and the coasts sighted. While Spanish and Portuguese official cartography only depicts America up to 35° south latitude, the maps of Waldseemüller (1507), Rosselli (1508) and Glareanus (1513) interpreted the third voyage by drafting the American coastline with a southwest bearing up to about 50° south latitude, that is, accepting the latitude stated by Vespucci (52° south latitude) but disregarding the stated bearing (southeast). If the expedition's purpose was to find a passage to the west, the course toward the southwest beyond the boundaries fixed by the Tordesillas Treaty (32° south latitude) would have been deliberately concealed by the Portuguese Royal Court to avoid diplomatic disputes with Spain, as it involved exploring into the Spanish hemisphere.
13. Lois (2008a: 257) remarked that "this way of trimming off the surface to be represented on maps was the first strategy to leave empty expanses in maps; there were no empty spots within the map. They turned to flat maps, whose extent was previously estimated according to globe geometric principles (360° longitude and 180° latitude) and, in order to suggest that the world was broader than the one appearing on the map and that the whole world was still unknown, they mapped just part of it".
14. The *Lettera di Amerigo Vespucci delle isole nuovamente trovate in quarto suoi viaggi*, letter written by Vespucci to Piero Soderini, was included as appendix in the *Cosmographiae introductio*.
15. Francesco Rosselli, World Map, ca. 1508. Repository: Royal Museums Greenwich—National Maritime Museum, Digital.

16. Johannes Schöner, Globe (1515). Repository: Yale University Beinecke Digital Collections.
17. Henricus Glareanus, Southern Hemisphere, ca. 1513. Repository: John Carter Library Map Collection, Digital.
18. Schöner's explanatory text accompanying his 1515 map—*Luculentissima quaedan Terrae totius Descriptio*—refers to information provided to an agent of Welser Bank in Madeira by Portuguese seamen, who had purportedly navigated through a strait to the south of the known lands in America (Stevenson 1904; Tanzi 1985).
19. Johannes Schöner, Globus, 1520. Repository: Doria/National Library of Finland
20. Cristopher de Haro and Nuño Manuel (1514) and Juan Díaz de Solís (1515–1516) carried out the first explorations looking for a passage to the West in the Río de la Plata. The latter was commissioned in 1514 to sail the coast of Brazil to the south and upon reaching the end of the mainland, sail northeast or northwest up to "*las espaldas de Castilla del Oro*" (the shoulders of Castilla del Oro), that is to say, the coast of the Southern Sea discovered by Balboa (Fernández de Navarrete 1946).
21. Maximilianus Transylvanus' Account, Medina 1896, t 1: 264.
22. Cosmographer Rui Failero's estimates on longitude reckoned a lesser breadth of the Southern Sea and situated the Moluccas within the Spanish sphere, outlined by the Treaty of Tordesillas. Padrón (2010) states that, on Diego Ribeiro's 1529 chart of the world, the breadth of the Pacific Ocean has grown from Weldseemüller's 83° to 134° longitude. Thus, taking into account that Pedro and Jorge Reinel participated in the development of the expedition maps, the odds are that the itinerary Magellan presented to Charles V had been designed based on Reinel's 1519 map of the world, which depicts the known land up to the Río de la Plata (Cerezo Martínez 1994).
23. Maximilianus Transylvanus' Account, Medina 1896, t 1: 264. Antonio Pigafetta (1963: 59) noted in his account of Magellan's voyage that they had seen a strait on a map made by cosmographer Martin Behaim, kept in the Treasury of the King of Portugal. Some authors assert, instead, that Magellan had been guided by Schöner's 1515 map, on which a strait is laid down but at 40° or 45° south latitude, between the New World and a ring-shaped continent in the temperate southern zone (Martinic 2006).
24. According to Fernández de Navarrete (1946), Magellan's proposal to reach the East by sailing west "seemed ideal but dangerous, because the American continent (still not fully known) was thought to stretch from north to south, with no passage whatsoever to connect both seas". Lopo Homem's map of 1519, included in the Atlas Miller, would show this alternative, by representing a mass of land surrounding the Southern Pole and connecting America to the south of Asia. Repository: Bibliothèque Nationale de France, Departement des Cartes et Plans, Digital.
25. López de Gomara, Francisco, *Historia General de las Indias,* 1554. Repository: Hispanic Digital Library–National Library of Spain. Pigafetta (1963: 62) referred to this alternative as follows: "Should we fail to discover the strait to

sail from one sea to the other one, the captain general had determined that we would keep sailing southwards to 75° south latitude, where there is no night during summer, or at least it is very short, and where there is no daylight during winter".

26. Although Pigafetta developed a map to be attached to that voyage accounts which depicts the *Estrecho de los Patagones* o *Estrecho Patagónico* (Patagonian Strait), it is the *Carta Universal* or *Padrón Real de Turín* of 1523 (Great Universal Maritime Chart), attributed to Giovanni Vespucci, the first copy of the chart that registered the discovery (Martinic 1999: 18).
27. Maximilianus Transylvanus' Account, Medina 1896, t 1: 276
28. As Lois noted (200a: 280), during the Renaissance "the theoretical and practical knowledge are not two separate approaches but, interwoven, provide each other the necessary conditions to enable mutual development (and, thus, each of them cannot be understood in isolation)". Roa-de-la-Carrera (2002) referred to the challenging of classical and medieval knowledge in his review of Vespucci's ideas and focus on the communication between theory and practice as a condition for the production of new cartographic knowledge and as a requisite for imperial expansion. Camprubi Bueno (2010) referred to the importance of Iberian States' discovery voyages and royal patronages in the development of theoretical geography and particularly, how the first circumnavigation financed by Charles V enabled the theory of the Earth's sphericity to reach a purely operational phase.
29. Captain Francisco de Hoces, who led one of the vessels of García Jofré de Loaisa Expedition, which was blown to the southeast by a gale from the eastern mouth of the strait, presumably to 56° south latitude, mentions in his 1526 account "the edge of the land" (Martinic 1911: 11).
30. O. Fine *Noua, Et Integra Uniuersi Orbis Descriptio*, 1531. Repository: Library of Congress. Geography and Map Division, Digital.
31. According to Lois (2008a: 101), the geographical hypothesis of a southern hemisphere might have been boosted decisively since Copernicus's *De revolutionibus* (1543), who restored the issue of the earth globe's balance and, "with the purpose of reinstalling an equilibrium – not only for aesthetics but also to recover an equilibrium as regards physics - a huge southern continent was brought back to the maps during the second half of the 16th century and it became bigger over time."
32. Gerardus Mercator, *Sphères terrestre et celeste*, 1541. Repository: National Library of Australia, Digital Collections.
33. Gerardus Mercator, *Planisfhere aux latitudes croissantes,* 1569 Repository: Bibliothèque Nationale - Departament des Cartes et Plans, Digital.
34. "*Primeramente vos doy licencia y facultad a vos el dicho Pero Sancho de Hoz para que por nos y en nuestro nombre y de la corona real de Castilla podáis navegar con los dichos navíos que ansí os ofrecéis a hacer por la dicha Mar del Sur, donde tienen las dichas sus gobernaciones los dichos Marqués don Francisco Pizarro y Adelantado don Diego de Almagro y don Pedro de Mendoza y Francisco de Camargo hasta el dicho Estrecho de Magallanes y la tierra que*

está de la otra parte del dicho Estrecho, sin que entréis en los límites y paraje de las islas de tierra que están dadas en gobernación a otras personas a conquistar ni gobernar ni rescatar […]", in Medina, t.1, 1896.

35. A. de Santa Cruz, *Islario general de todas las islas del mundo*, c. 1545. Repository: Hispanic Digital Library—National Library of Spain.
36. A. de Santa Cruz, *Tierra o isla al mediodía del estrecho de Magallanes—Islario*, c. 1545. Repository: Hispanic Digital Library – National Library of Spain.
37. Alonso de Santa Cruz, Map of the World, 1542. Repository: National Library of Australia.
38. According to Tanzi (1985), before sailing into the strait, one of the vessels would have been driven by a gale to the south, in the Atlantic Ocean.
39. B. Agnese, *Atlas*, 1544. Repository: Hispanic Digital Library – National Library of Spain.
40. G. Gastaldi, *Universale,* 1546. Repository: John Carter Brown Map Collection.
41. G. Gastaldi, *Carta Marina Nova tabula*, 1548. Repository: David Rumsey Map Collection, David Rumsey Map Center, Stanford Libraries.
42. Gastaldi's maps and those inspired by his work (G. Ruscelli, P. Forlani, F. Bertelli, G. F. Camocio among others) do not use the name *Terra Australis. They repeat three toponyms*: *Tierra del Fuego*, *Tierra del Fuego Incognita* and *Terra Incognita,*
43. F. Bertelli, *Universale descrittione di tutta la terra conosciuta fin qui*, 1565. Repository: Library of Congress Geography and Map Division Washington, D.C.
44. G. Ruscelli, *Orbis Descriptio. Tavola Universal […],* 1561. Repository: David Rumsey Map Collection, David Rumsey Map Center, Stanford Libraries
45. Maps of Reinel (1522), Salviati (1525), Vespucio (1526), Ribero (1529), Santa Cruz (1542), Caboto (1544), Agnese (1545).
46. Two regions are identified on the maps depicting the southern continent between 1515 and 1540: *Brasilia regio* and *Regio Patalis.* The first one appears on Schöner's 1515 and 1520 globes and is seen to the south of the strait that separates the austral continent from the New World, for which it can be a displacement or confusion with the territory of Brazil. The toponym *Patalis Regio* would derive from one of Plinio's writings referring to a port in India to the south of the Equator. Afterwards, Roger Bacon locates it to the south of Tropic of Capricorn and Pierre D'Ailly keeps the reference in his *Imago Mundi.* It is very likely that this last work might have served as Schöner and Fineus' inspiration (Van Duzer 2007).
47. Portuguese Jorge de Meneses' voyages to New Guinea coasts (1526) and those of Spanish Alvaro de Saavedra (1527–1528) and Ruy López de Velasco (1542–1545) and the French expedition to Sumatra led by the Parmentier brothers (1529–1530).
48. Diego Ribero, *Carta universal*, 1529. Repository: National Library of Australia Digital Collections

49. Nicholas Vallard (1547), Pierre Desceliers (1546), Gulillaume Le Testu (1555). Repositories: National Library of Australia, Digital; The Huntington Library, California; Bibliotèque Nationale de France, Digital.
50. Guillaume Le Testu, *Cosmographie Universelle,* 1555. Repository: Bibliothèque Nationale de France, Départament des Cartes et Plans, Digital.
51. Wytfliet, C. *Chica sive Patagonica et Australis Terra,* 1597. Repository: David Rumsey Map Collection, David Rumsey Map Center, Stanford Libraries.
52. We thank Professor Ignacio Antelo Reinoso for the translation of Wytfliet's Latin text and his comments on possible meanings of the name Chica.
53. F. García Jofré de Loaysa and J. S. Elcano (1525–1526), S. de Alcazaba (1534–1535) and navy of the Bishop of Plasencia's expeditions (1539-1540).
54. Gonzalo Fernández de Oviedo, *Historia General y Natural de las Indias* [1535–1557], cited in Nocetti and Mir (1997)
55. Claudio Gay, *Historia Física y Política de Chile. Documentos sobre la historia, la estadística y la geografía de Chile*, Tomo I, París, 1846, cited in Nocetti and Mir (1997)
56. Capitulation and filing by Pero Sancho de Hoz to discover places within Southern Sea, in Medina 1901, t. VIII.
57. Diego Gutiérrez, *Americae sive qvartae orbis partis nova et exactissima descriptio,* 1562. Repository: Library of Congress. Geography and Map Division, Digital
58. Juan López de Velasco, *Geografía y descripción universal de las Indias*, 1575. Repository: The Huntington Library, San Marino, California.
59. Abraham Ortelius, *Typus Orbis Terrarum,* 1570. Repository: Library of Congress. Geography and Map Division, - Digital
60. H. L. V. de la Popelinière, *Les Trois Mondes* (1582). Repository: John Carter Brown Library, Brown University, Digital.
61. Andrews (1968) and Spate (2004) deduced the intent of looking for *Terra Australis* from the route followed by Drake.
62. Drake could have used the 1570 map of cartographer Fernando Vaz Dourado who outlines the strait with an open "V" orientation (NESW-NW), sketching to the south a cluster of islands and a coastline which is interrupted at about 60º south latitude. Repository: The Huntington Library, San Marino, California.
63. Joan Martines, *Southern South America, Portolan Atlas,* 1578. Repository: The Huntington Library, San Marino, California.
64. News of Drake's discovery has reached England in different ways. According to Jesuit José de Acosta (1590), pilot Nuño da Silva, when being questioned by New Spain Viceroy in 1579, had declared that "*the Strait became an island and both seas come together*". Likewise, Spanish Ambassador in England, Bernardino de Mendoza, would have reported it to the Spanish Crown in 1582, after receiving information from a person who would have seen Drake's letter (Spate 2004).
65. Abraham Ortelius, *Maris Pacifici,* 1589. Repository: David Rumsey Map Collection, David Rumsey Map Center, Stanford Libraries.

66. Jodocus Hondius, *Americae novissima Descriptio,* 1589. Repository: Bibliothèque Nationale de France, Département des Cartes et Plans, Digital.

References

Acosta, J. (1590) *Historia natural y moral de las Indias.* Memoria Chilena – Biblioteca Nacional Digital de Chile

Almagiá, R. (1939), "Intorno ad un grande mappamondo perduto di Giacomo Gastaldi (1561), *La Bibliofilia* (41) 7/8: 259–266

Andrews KR (1968) The Aims of Drake's Expedition of 1577-1580. The American Historical Review 73(3):724–741

Besse JM (2010) "Philologie et cartographie au XVIe siècle: la critique de l'histoire dans le Parergon et le Thesaurus geographicus d'Abrabam Ortelius", *3°*. Agendas para História da Cartografia Iberoamericana, Universidade de São Paulo, Brazil, Simpósio Iberoamericano de História da Cartografía

Buisseret, D. (2003) *La revolución cartográfica en Europa. 1400–1800. La representación de los nuevos mundos en la Europa del Renacimiento,* Barcelona, Paidós

Callander J (1967) *Terra Australis cognita or voyages to the Terra Australis* [facsimile del original, Edimburgo, 1766]. Ámsterdam 1:110–201

Camprubí Bueno, L. (2010) "Viaje alrededor del Imperio: rutas oceánicas, la esfera y los orígenes atlánticos de la revolución científica", *El Catoblepas* (95)

Cattaneo A (2009) Réflexion sur les climats et les zones face á l'expansion des XVe et XVIe siècles. Le Monde des cartes 199:7–21

Cerezo Martínez R (1994) *La cartografía marítima española en los siglos XIV*. XV y XVI, Madrid, Consejo Superior de Investigaciones Científicas

Crone, G. R. (1946) "A manuscript Atlas by Battista Agnese in the Society's Collection", *The Geographical Journal* (108) 1/3: 72–80

Chartier, R. (1996) *El mundo como representación. Historia cultural: entre práctica y representación,* Barcelona, Gedisa

Chaunu P (1984) *Conquista y explotación de los nuevos mundos (siglo XVI).* Labor, Barcelona

Day, J. T. (1988) *The New World in Maps: The First Hundred Years,* The Newberry Library

Fernández de Navarrete M (1946) *Colección de los viajes y descubrimientos que hicieron por mar los españoles.* Buenos Aires, Guarania

Fischer J, von Wieser F (1907) *Cosmographiae introduction of Martin Waldseemüller*. United States Catholic Historical Society, IV, New York

Foucault M (1998) Las palabras y las cosas. Una arqueología de las ciencias humanas, Madrid, Siglo Veintiuno editores

Grande S (1905) *Le carte d'America di Giacomo Gastaldi.* Harvard University, Digital, Repositorio

Giucci G (2014) *Tierra del Fuego: la creación del fin del mundo.* Buenos Aires, Fondo de Cultura Económica

Haesbaert R, Limonad E (2007) El territorio en tiempos de globalización. Revista Electrónica de Ciencias Sociales Aplicadas 1–2:39–52

Harley JB (2005) *La nueva naturaleza de los mapas.* Ensayos sobre la historia de la cartografía, México, Fondo de Cultura Económica

Hiatt A (2007) "The Map of Macrobius before 1100", *Imago Mundi.* The International Journal for the History of Cartography 59(2):149–176

Kagan, R. L. (2005) "La Luna de España: Mapas, ciencia y poder en la época de los Austrias", *Pedralbes,* 25

King RJ (2009) The Jagiellonian Globe: A key to the puzzle of Java la Grande. The Globe 62:1–50

Lestringant, F. (2015) La Cosmographie Universelle de Guillaume Le Testu (1556), de l'Asie des merveilles aux Nouveaux Mondes en construction, París

Levillier, R (1956) "Mundus Novus", *Anales de la Universidad de Chile*, 104, Santiago de Chile

Lois, C. (2008a). *Plus Ultra Equinoctialem. El descubrimiento del hemisferio sur en mapas y libros de ciencia en el Renacimiento.* Unpublished doctoral thesis. Facultad de Filosofía y Letras, Universidad de Buenos Aires

Lois C (2008b) *América quarta pars*: ¿isla o continente? El debate conceptual sobre el estatus geográfico del Nuevo Mundo en el siglo XVI. Fronteras de la Historia, Instituto Colombiano de Antropología e Historia 13(2):259–279

Lois C (2010) "*¿Quinta Pars o Terrae Incognitae*? La cuestión de la verosimilitud en la representación cartográfica de lo desconocido", *3°*. Agendas para História da Cartografia Ibernoamericana, Universidad de São Paulo, Brazil, Simpósio Iberoamericano de História da Cartografía

López de Gomara, F. (1554) *Historia General de las Indias,* Biblioteca Digital Hispánica – Biblioteca Nacional de España

López de Velasco, J. (1575) *Geografía y descripción universal de las Indias,* University of California Libraries – Digital

Luiz, M. T. and Schillat, M. (1997) *La frontera austral. Tierra del Fuego 1520–1920,* Cádiz, Servicio de Publicaciones de la Universidad de Cádiz

Martinic BM (1981) Estrecho de Magallanes, territorio marítimo chileno. Anales del Instituto de la Patagonia, Punta Arenas, p 12

Martinic (1998) "Drake y el descubrimiento de la insularidad fueguina. La evidencia cartográfica", Anales del Instituto de la Patagonia, Punta Arenas, 26

Martinic BM (1999) *Cartografía Magallánica 1523-1945.* Ediciones de la Universidad de Magallanes, Punta Arenas

Martinic BM (2006) Historia de la Región Magallánica. Ediciones de la Universidad de Magallanes, Punta Arenas

Medina JT (compilator) (1896) Colección de documentos inéditos para la historia de Chile, t. 1, 1558–1572, Santiago de Chile, Fondo Histórico y Bibliográfico J. T. Medina

Medina (1901) Colección de documentos inéditos para la historia de Chile, t. 8, 1518–1818, desde el viaje de Magallanes hasta la batalla de Maipo, Santiago de Chile, Fondo Histórico y Bibliográfico J. T. Medina

Mignolo W (1995) *The darker side of the Renaissance: literacy, territoriality and colonization.* University of Michigan Press, Ann Arbor

Nocetti, O. R. and Mir, L. B. (1997) *La disputa por la tierra. Tucumán, Río de la Plata y Chile (1531–1822),* Buenos Aires, Sudamericana

O´Gorman, E. E (1995) *La invención de América.* México, Fondo de Cultura Económica

Padrón R (2002) Charting Empire, Charting Difference: Gomara's Historia General de las Indias and Spanish Maritime Cartography. Colonial Latin American Review 11(1):47–69

Padrón R (2004) *The Spacious Word: Cartography, Literature, and Empire in Early Modern Spain.* University of Chicago Press, Chicago

Padrón R (2009) A Sea of Denial: The Early Modern Spanish Invention of the Pacific Rim. Hispanic Review 77(1):1–27

Padrón R (2010) "Las Indias olvidadas: Filipinas y América en la cartografía imperial española", *3°*. Universidade de São Paulo, Brazil, Simposio Iberoamericano de História da Cartografía

Pigafetta A (1963) *Primer viaje en torno del globo.* Espasa-Calpe, Madrid

Portuondo, M. (2013) *Ciencia secreta: la cosmografía española y el Nuevo Mundo,* Madrid, Iberoamericana

Quintero S (2000) Pensar los mapas. Notas para una discusión sobre los usos de la cartografía en la investigación social. Escolar, C. (compilator), *Topografías de la investigación. Métodos, espacios y prácticas profesionales,* Buenos Aires, Eudeba

Raffestin C (1993) *Por uma Geografia do Poder.* São Paulo, Atica

Roa-de-la-Carrera C (2002) El Nuevo Mundo como problema de conocimiento y el discurso geográfico del siglo XVI. Hispanic Review 70(4):557–580

Rodamilans Ramos, F. (2010) "Crónica de Pedro Fernández de Quirós: Historia del descubrimiento de las regiones australes hecho por el General Pedro Fernández de Quirós", *Ab Initio*, 1

Romano R, Tenente A (1979) *Los fundamentos del mundo moderno.* Edad Media tardía, Renacimiento, Reforma, Madrid, Siglo XXI

Sánchez A (2010) "La voz de los artesanos en el Renacimiento científico: cosmógrafos y cartógrafos en el preludio de la 'Nueva Filosofía Natural'", *Arbor*. Ciencia, Pensamiento y Cultura 186(746):449–460

Sankey M (2003) The French and Terra Australis. The Journal of the Sydney University 25:26–63

Santa Cruz, A. de (1545) *Islario general de todas las islas del mundo,* Biblioteca Digital Hispánica – Biblioteca Nacional de España

Spate O (2004) *The Spanish Lake: The Pacific since Magellan.* ANU Press, Australia

Stevenson EL (1904) Martin Waldseemuller and the Early Lusitano-Germanic Cartography of the New Word Author(s). Bulletin of the American Geographical Society 36(4):193–215

Stevenson EL (1921) *Terrestrial and celestial globes: their history and construction, including a consideration of their value as aids in the study of geography and astronomy.* Hispanic Society of America by the Yale University Press, New Haven

Tanzi HJ (1985) El continente antártico y la Tierra del Fuego. Revista de Historia de América 100:13–54

Van Duzer Ch. (2007) Cartographic Invention: The Southern Continent on Vatican MS Urb Lat. 274, Folios 73v-74r (c. 1530)", *Imago Mundi. The International Journal for the History of Cartography*, (59) 2: 193–222

Van Duzer Ch (2008) "A Newly Discovered Fourth Exemplar of Francesco Rosselli's Oval Planisphere of ca. 1508", *Imago Mundi. The International Journal for History of Cartography*, (60) 2: 195–201

Vignolo P (2003) Nuevo Mundo: ¿un mundo al revés? Las antípodas en el imaginario del Renacimiento. Bonnett, D. and Castañeda, F. (ed.), *El Nuevo Mundo. Problemas y debates*, Bogotá, Uniandes

Wagner HR (1931) The manuscript atlases of Battista Agnese, Papers of the Bibliographical Society of. America 25(1):1–110

Wallis H (1984) The Cartography of Drake's voyage. In: Thrower NW (ed) *Sir Francis Drake and the famous voyage*. Los Angeles, University of California, pp 1577–1589

Woodward D (2007) *The Italian map trade, 1480–1650*. Woodward D. (ed), *The history of cartography*, Vol. 3, University of Chicago Press

Zuber MA (2011) The Armchair Discovery of the Unknown Southern Continent: Gerardus Mercator, Philosophical Pretensions and a Competitive Trade. Early Science and Medicine 16(6):505–541

Chapter 3
Cartographic Construction of a South Atlantic Region

Abstract The study of cartography offers valuable information on the territorial genesis of the southern tip of America and enables us to observe the transformation of a space perceived as *finisterre* or border of the ecumene into a territory shaped on converging economic and strategic interests of different actors. We explore the cartographical distinction of a South Atlantic Ocean region in the context of the competition for the use of navigable routes and, from the end of the 18th century, for the marine-faunal resources. We consider the role of map as a key instrument for keeping certain dominion over the southern space and we explore how the cartographic discourse works in the valorization of the territory, provides a set of references that identifies it and conditions the interventions' nature and scope.

Keywords Territorial distinction · Geopolitical region · Southern Patagonia · Tierra Del Fuego and South Atlantic Ocean · Cartography · 16th, 17th and 18th centuries

3.1 Introduction

During the process of discovering and exploring the southern hemisphere, cartographic representation of the southern tip of America adopts several configurations that result not only from the knowledge of geographical development. It also derives from interpretations of its territorial nature, as well as from map producers and promoters' interests. Maps not only describe a territory but also produce it with an underlying intent. They "propose a space to act" unveiling possibilities and anticipating effects (Kitchin and Perkins 2009).[1]

Bearing in mind the propositional role of maps, we analyze the differentiating process of a southern Atlantic Ocean region from the 16th century to 18th century, in the context of the competition for the use of navigable routes and the access to marine resources of growing commercial interest.

Cartography—the condition and result of territorialization processes—constitutes a key instrument to keep certain political, scientific, functional and symbolic

L. I. de Lasa and M. T. Luiz, *The Southernmost End of South America Through Cartography*, The Latin American Studies Book Series,
https://doi.org/10.1007/978-3-030-65879-3_3

dominion over the represented space. Likewise, it takes part in the process of valorization and significance of space, providing a set of references over which European knowledge, beliefs and expectations build a territorial identity.

We consider the first territorial narratives following the discovery of the route of the Magellan Strait and the genesis of an imaginary that linked adjacent lands to the south with an alleged southern continent, conceived as a "fifth part" of the orb. We observe the transformation of an unexplored space, perceived as the border of the ecumene into a territory incorporated to the imperial designs, thought as a geopolitical region upon the convergent interests of different States.

The interpretation of this process rests on the identification of sets of maps that present different configurations, proposals of space frames, data selection and toponym inscription, which are decisions taken in the light of imperial expansion goals and projects. Some representations, reflected in long-term cartographic conceptions, shape geographic imaginaries and play a key role in the legitimation of geopolitical intentions.

Following the criteria of sources systematization, we first introduce the 16th century maps that show an unknown southern hemisphere. The said maps represent lands and seas from empirical knowledge and trade expansion expectations. Secondly, we explore maps produced during the last third of the 16th century and the first decades of the 17th century, which focus on the Pacific Ocean as a new field of action and put forward the first differentiation of the Magellan territory as a strategic area to connect with Asia. Lastly, in 18th century cartography, we note that the interest shifts to the South Atlantic Ocean and the inclusion of the Malvinas Islands in the plans of occupation due to its crucial position in interoceanic communication and the abundance of marine-fauna resources of growing commercial interest.

3.2 Finding the Southwestern Pass and the First Territorial Narratives

The first representations of the southern hemisphere describe the scope of geographic imagination within the debate on the *imago mundi* and the ideas about the existence of lands or islands of identical nature to America that is habitable. The hypothesis of a huge continent, incorporated to the orb as a "fifth part", gained credibility among northern Europe cartographers and provided a stimulus to expansion through the attractive manuscript or printed maps. Conversely, kingdoms that were early involved in overseas endeavors designed their strategies on maps that outlined lands and islands empirically confirmed. Such manuscript maps, produced by cartographers in the service of Spain and Portugal, constituted a necessary tool for the development of exploration and conquest.

Both empires developed a similar cartographic policy, centralizing information and map production registers in institutions that directly depended on the Crown. The *Padron Real*, official map elaborated and updated in the House of Trade in

Seville, was an important territorializing instrument that represented monarchical dominion spaces, integrating effective and symbolic possessions (Sánchez 2017). Another element shared by Iberian Empires is the secrecy policy, in order to guarantee the monopoly of knowledge. Measures implemented to assure confidentiality failed to prevent the circulation of relevant charts and data through espionage—a common practice in Spanish–Portuguese relationships—and through the interaction of captains, pilots, cartographers and cosmographers temporarily hired by both Crowns (Destombes 1955; Porro Gutiérrez 2003; Harley 2005). The differences stem from their territorial occupation modalities. The Spanish aimed to represent territories for the purpose of colonialization. The Lusitanians, in turn, opted for a system of commercial enclaves and focused on controlling maritime routes, except in Brazil that was well mapped since it was a colonial settlement (Buisseret 2007).

Maps of the southern hemisphere that represent lands up to latitude 50° S reflect the territorial controversy among Iberian monarchies over the manipulation of geographical data. The Portuguese, who attempted to prevent other European Crowns to vie for eastern commerce by controlling the route through the Cape of Good Hope, showed arbitrarily longer routes in order to inflate expenses of the trip to India. In turn, the Spaniards opted to diminish the width of the Pacific Ocean with the purpose of representing a shorter route, locating the Spice Islands within their area of influence, delimited in the Treaty of Tordesillas.[2] This is also present on all maps based on the *Padrón Real* and 1519 Pedro and Jorge Reinel's map of the world, possibly used by Magellan to design the course of the voyage proposed to Charles V (Fig. 3.1).[3]

This map of the world enlightens the geopolitical situation in a time of great dynamism regarding overseas expansion. Focusing on the line of Tordesillas, flags indicate Portuguese and Castilian jurisdictions and ships show the African route and the search of a possible passage to the East round the north or the south of the New

Fig. 3.1 Reinel Planisferio (1519)

World, situating the latter in the Río de la Plata, according to the information provided by Juan Díaz de Solís (1516). In addition, this map introduces elements that could have favored the royal decision to sponsor the Magellan venture. By representing the South Sea with a smaller size, based on cartographer Rui Faleiro's longitude calculations, the image shows the feasibility of reaching the Moluccas through the hypothetical southern route, as well as from the coasts of Panama.

Following the first reconnaissance of the entrance to the Magellan Strait, pilot Esteban Gómez, with the support of other officials, captured captain Alvaro de Mezquita. He then controlled the ship San Antonio and deserted the Magellan expedition. Although his reasons are unclear, he probably received the order to communicate the existence of the southwestern passage to ship-owner Cristóbal de Haro, who concurrently planned another expedition to the spice producers from Panama. Pedro Reinel elaborated, between 1522 and 1524, a map of the southern hemisphere with a polar projection, which represented the Patagonian coastline up to 52° south latitude and a deep bay that suggested a possible path, although it is not identified as a strait or as a channel (Martinic 1999).[4] This chart also locates an island in the place of the Malvinas, probably based on the map of Andrés de San Martín, who was the pilot of the San Antonio ship during the reconnaissance performed from San Julián between July and August 1520, before reaching the Magellan Strait (Destombes 1955; Cerezo Martínez 1994).

The map of the southern tip of America made by chronicler Antonio de Pigafetta, published with the report on Magellan-Elcano voyage (ca. 1525), offers the first depiction of the *Streto patagónico* as a straight channel that connects the *Mare Oceano* with the *Mare Pacifico*. However, the 1523 *Carta Universal* or *Padrón Real* of Turin, attributed to Juan Vespucci, is the first world map that includes the discovery of the southwestern path. Such a depiction of the Magellan Strait, schematic but conceptually correct, was kept in official Spanish cartography and contributed to the representation of Tierra del Fuego as part of the vast southern continent, an idea that persisted well into the beginning of the 17th century.

After Magellan's death, the expedition reached Tidore in November 1521. Exhausted due to the long journey and weakened because of Elcano's return to Spain, the Portuguese easily neutralized it (Chaunu 1984). Together with the military and commercial rivalry, a diplomatic dispute had arisen over the dominion of Moluccas and required the intervention of boards of experts at Badajoz-Elvas. The Treaty of Zaragoza in 1529 put an end to the controversy, with Spain waiving its rights to the islands that belonged to the Portuguese Crown. Such dispute gave rise to a cartographic distortion of information, although, as Spate (1979) remarked, the Portuguese were closer to the truth since the antemeridian of the line of Tordesillas (134° 40′ E) is seven degrees to the East of the Spice Islands. Manuscript maps of Salviati (1523), Vespucci (1526), Maiolo (1527) and Ribero (1529) provide a similar configuration of the Magellan Strait. They gradually unveil the coasts of Tierra del Fuego from the northern tip toward the south, fully depicting the Pacific Ocean and locating the Spice Islands within the Spanish hemisphere.

Maps representing the alleged southern landmass base themselves on islands, archipelagos and coastal features information that, construed in the light of classical and medieval sources, enable us to outline a constant coast that ran alongside the 50° south latitude parallel, except for two big peninsulas that reached the Equator.

Tierra del Fuego cartographically revitalized the *Terra Australis* (Parry, in: Spate 1979) and the elaboration of new proposals expressed the scientific and political expectations. The first printed map representing the Magellan Strait, produced in 1531 by Oronce Fine, incorporates the outline of the adjacent land—to the south—to the continent named *Terra Australis* (Martinic 1999; Sankey 2003).[5] By interpreting the information on the coasts of Tierra del Fuego and New Guinea on the same premise, Gerardus Mercator concluded the construction process of a new cartographical image of the world, by depicting on his 1541 globe a huge southern continent. This configuration appeared again in his 1569 map of the world.[6]

Concurrently, there were other hypotheses that proposed an insular configuration of southern lands, ranging from island, archipelago or *isla de tierra*. The latter, mentioned in the *Capitulación y asiento que se tomó con Pedro Sancho de Hoz para efectuar descubrimientos en la Mar del Sur*, signed on 24 January 1539, evinces that such high expectations toward novelty boosted geographical imagination beyond already known terrestrial categories within the current classifications. These categories were continent, island, peninsula and isthmus. As Lois remarked (2018: 57), "uncertain geographies underwent several rehearsals in which such concepts were applied, and they could be redefined using any new combination of such concepts in case the former formula resulted unsatisfactory". The invention of the concept "*isla de tierra*"[7] could have resulted from the necessity to define the specific nature of islands of big size, such as Tierra del Fuego, as described by Alonso de Santa Cruz in his *Islario General* (ca. 1542 and 1545).[8] The cosmographer of the House of Trade states that the land to the south of the Strait, "though not fully or accurately discovered, it is known for a fact to be an island, the biggest in the world, which is the nearest land to the Antarctic Pole". Nevertheless, the cartographic representation left some room for doubt by prolonging longitudinally the eastern coast of Tierra del Fuego up to the end of the map and by including the legend *Tierra o isla del estrecho de Magallanes* ["Land or island of the Magellan Strait"].[9]

Santa Cruz' work also offers data that led to consider an early discovery of the Malvinas Islands. Upon describing the itinerary of the Magellan expedition prior to reaching the Strait, he referred to what one of the ships discovered, by stating: "they sailed ahead, having reached and encountered several islands located eighteen leagues to the east of San Julián Port, which they named Sanson and de Patos [...] and are fifty-one degrees of height". This last piece of information is confusing since the map locates these islands at 49° south latitude, opposite to what today is Puerto Deseado. It also depicts two other nameless islands between 51° and 53° south latitude, near a bay formed by two peninsulas of the northern coast of the Land or Island of the Magellan Strait. Such islands would coincide—considering the latitude—with the location of the Malvinas.

Another statement in *Islario* enabled the attribution of the discovery to one of the ships of Francisco de Camargo's expedition that allegedly stayed at the archipelago during ten months of 1540, before returning to Spain:

> [...] the Bishop of Plasencia's army is the one who has discovered most of this part or island dealt with today [Tierra del Fuego] who, before entering the Strait, discovered over one hundred leagues of coast. Very high land, full of mountains and inhabited, since they were asked to light fires. And to the south of the Strait, they discovered over sixty leagues of coastline and two big islands with an archipelago of smaller ones by which they spent several days due to lack of time to conquer them.

Santa Cruz, *Islas del Estrecho de Magallanes*, from the *Islario general de todas las islas del mundo*, c. 1545, page 709

http://bdh-rd.bne.es/viewer.vm?id=0000149359&page=1

Both the conceptual ambiguity, manifest in the wording of "land or island" or "part or island",[10] and the dubious correspondence between texts and maps, could be considered ways of expressing—intentionally or unintentionally—the temporary status of interpretations. However, Santa Cruz' proposal on the insular nature of southern lands will prove to be closer to the geographical reality of Tierra del Fuego than that of his influential contemporaries. On the one hand, he explicitly criticizes "Oroncio [Fine] and others out of Spain", who supported the conception of a big landmass and, on the other hand, gives credit to the information provided by the expedition of Francisco de la Rivera (Bishop of Plasencia's navy) as regards to the discovery of "two huge islands within an archipelago of tiny ones".

André Thevet's island book (1586) also reinforces the hypothesis of the discovery of the Malvinas by the San Antonio vessel from Magellan's expedition. *Le grand insulaire et pilotage* includes three charts that provide more details of the islands of the southern end of America, represented on former maps used as a source, Santa Cruz (1545) and Mercator (1569) among others. *Les Isles Menues* represents Pacific islands from 50° south latitude to the entrance of the Magellan Strait (52° south latitude) and the coasts of *Terre de Feuz* up to 56° south latitude. *Petit Arcipelage* comprised islands located to the west of Tierra del Fuego and contiguous to *Terra Australis.* The third chart, *Les Isles de Sanson ou des Geantz,* represents islands in *Mer Australle* to the east of Patagonia, whose configuration and location in latitude—49° and 50° south latitude—makes us guess they are the Malvinas.[11] In addition, the cartographer mentions having received maps from a Portuguese captain. Therefore, it is very likely that Álvaro da Mesquita, the Captain of the San Antonio ship, met Thevet in Lisbon, presumably between 1563 and 1567, and gave him various maps. Among them, the chart of the archipelago resulted from the reconnaissance performed between July and August 1520 by the pilot Andrés de San Martín (Laguarda Trías 1994) (Fig. 3.2).

Following the sighting registered by Sebald de Weert in 1599, European cartography incorporated the Malvinas and named them after him, Sebald Islands or Sebaldines. Such names persisted until the 18th century, together with the name New Islands given by Saint Malo traders, and Falkland, derived from the name given

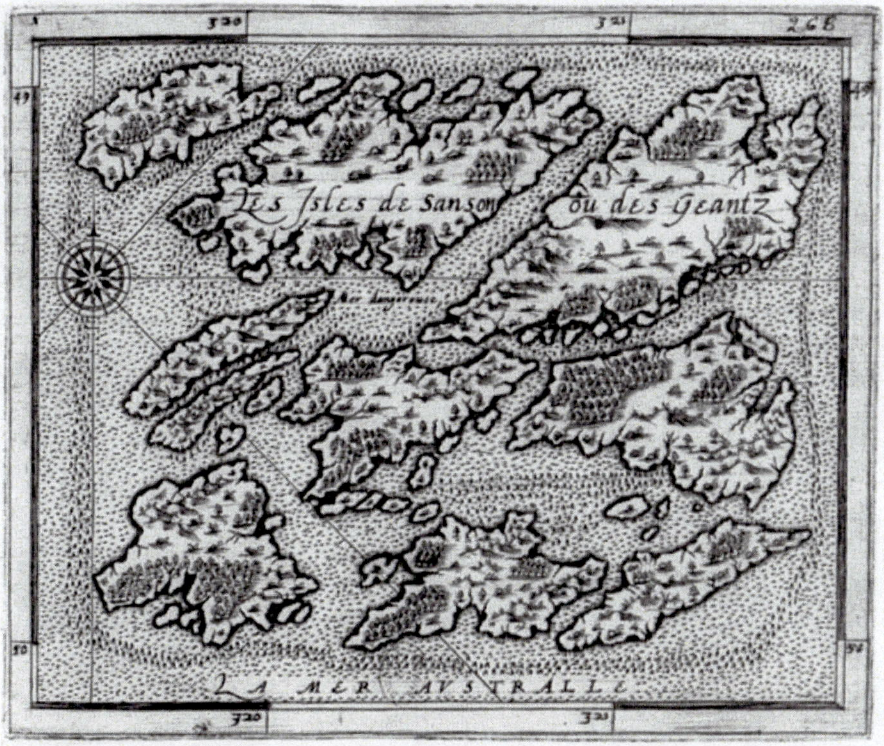

Fig. 3.2 Thevet *Les Isles de Sanson ou des Geantz* (ca. 1586)

by John Strong (1690) to the strait that separates the two bigger islands within the archipelago.

3.3 The Process of Territorial Distinction

Cartographical sources of the second half of the 16th century evinced the attention paid by Spain to the Pacific Ocean from a two-fold interest: to guarantee the access to the islands in the south of Asia and to find new temperate southern lands. The conquest of the Philippines (Legazpi) and the finding of the return route from the Moluccas to New Spain across the northern hemisphere (Urdaneta) in 1565, and two years later, the discovery of the Solomon Islands (Mendaña), fueled such expectations. These voyages were paramount to the expansion toward the Pacific Ocean and cartography mirrored the territorialization process of the new ocean as a space of Spanish dominion (Schurz 1922; Spate 1979).

The manuscript map of Juan López de Velasco, first Cosmographer and Chief Chronicler of the Council of the Indies, was included in his *Geografía y Descripción universal de las Indias* (1575) and published in Antonio de Herrera's *Descripción de las Indias Occidentales* (1601).[12] It offers the official image of overseas Hispanic possessions by means of a trans-Pacific Spanish hemisphere demarcated by the Treaties of Tordesillas and Zaragoza. The depiction shows the territories of the intertropical zone—Viceroyalties of New Spain and Peru, Philippines and Solomon Islands—in the center of the Pacific Ocean, the coasts of New Guinea and Tierra del Fuego remain open, leaving considerable space for the existence of southern temperate lands. From the "*Indias del Poniente*" (East Indies) to the "*Indias del Mediodía*" (South America), navigation and trading routes cross the Pacific Ocean between Acapulco and Manila (Philippines). This was a privileged connection within the imperial strategy, since reaching the South Sea through the southwestern route resulted in a more costly alternative due to the distance and the adverse navigability conditions in the Strait (Fig. 3.3).

Battista Agnese's manuscript maps of America and the Pacific Ocean are the antecedent of this layout that draws away from the maps reproducing the *Ptolomeic* notion of a tripartite *orbis terrarum*, adding a fourth world surrounded by an ocean

Fig. 3.3 López de Velasco *Demarcación y navegaciones de Yndias* (1575)

to the left (Padrón 2009). The Venetian cartographer, hired by Charles V to produce his atlas between 1540 and 1542, anticipates the image focused on the Pacific Ocean and the geographical selection of López de Velasco. Thus, he indicates the imperial interest spots in the first half of the 16th century: China, the Moluccas and New Guinea. The Magellan Strait, depicted as in the *Padrón Real*, is the only route to the Pacific Ocean and east–west direction of the ships shows its means of access.[13]

Agnese, *Perú* (1554)
https://collections.rmg.co.uk/collections/objects/541124.html

The anonymous map attributed to the Portuguese Antonio Pereira (1545) also represents the routes to the East, by placing a ship at the entrance of the Strait, in the Atlantic Ocean, and in the Pacific Ocean other four ships sailing in a straight line toward New Guinea. To the north of the Equator, two other ships sailing from Mexico indicate the connection with the Moluccas.[14] Like Santa Cruz' island book and Cabot's world map, it incorporates data on the Bishop of Plasencia's Navy exploration and possibly on other non-official expeditions and presents a peculiar configuration of the tip of the continent. It outlines the northern coastline of Tierra del Fuego toward the Atlantic and the Pacific oceans and depicts a set of islands and channels open to the south. Homen Reinel's map of the world (1559)[15] and Domingo Texeira's map of the world (1573)[16] keep this layout, whereas Vaz Dourado's maps of the Magellan Strait (1568, 1570) show it as an "island-archipelago" (Fig. 3.4).[17]

Cartographer Joan Martines' 1578 Chart of South America, following the configuration of Ortelius' big southern landmass, represents Spanish possessions in the New World to the south of the Tropic of Capricorn. Contrary to what happens with López de Velasco's map, it confers a special visibility to the Magellan Strait and to Tierra del Fuego, by means of colors and shading of the coasts toward the Antarctic Circle and the inscription of a main toponym.[18]

Cartographical production, the work of chroniclers and cosmographers[19] and several documents of the colonial administration—specifically journeys reports and capitulations—show that toward mid-16th century, Spanish and English imperial strategies were considering hypotheses on the continental or insular nature of the lands contiguous to the Strait toward the Southern Pole. The former, posed by O. Fine (1531) and his followers, among others G. Mercator, depicted Tierra del Fuego as part of a vast southern continent. The latter offered two alternatives: Portuguese cartographers' proposal of an "island-archipelago" and the configuration of an island of big size (*isla de tierra*?), supported by Santa Cruz and depicted on B. Agnese's world maps of oval projection (1544)[20] and G. Gastaldi (1546 and 1548).[21]

The debate on the geographical reality of the southern hemisphere also nurtured the imaginaries—geographical and imperial—of the States that did not participate in the initial distribution of the New World. We are unsure of the intentions of the promoters of the first English trans-Atlantic initiatives. Nevertheless, it is likely that their project consisted of exploring the coasts of South America and of ascertaining

Fig. 3.4 Vaz Dourado *Atlas* (1570)

the natives' predisposition in view of the settlement of bases in the Río de la Plata, the Magellan Strait and into the Pacific Ocean, to the south of the Spanish settlements in Chile (Andrews 1968, 1982). Toward the end of the 1570 s, Queen Elizabeth and her advisors shifted their attention to the lands "beyond the equator", the lands between the Tropic of Capricorn and the southern tip of the continent, represented on Spanish, Portuguese and Flemish maps as a space devoid of European settlements (Andrews 1981; Padrón 2009).

The first English global strategy was designed bearing in mind two representations of the lands situated to the south of the interoceanic passage. According to Wallis (1984), Francis Drake would have taken a copy of Ortelius' map, published in 1564, together with another map especially produced by Vaz Dourado for the expedition. Although this world map was not found, others of the same cartographer (1570, 1572, 1576, 1580) show a southern region frame similar to that of Martines and represent Tierra del Fuego as an incomplete archipelago to the south.[22] Drake's itinerary along the shore of Tierra del Fuego in the Pacific Ocean enabled the confirmation of the insular status as it verified the absence of mainland.

The English Crown did not take part in the organization of the new expeditions, and those led by Cavendish (1586 and 1591), Chidley and Merrich (1589) and Hawkins (1593) had neither commercial nor territorial occupation purposes. However, Drake's presence initiated an English series of navigations through the Strait in order to explore, trade and plunder (Bradley 1989). Later privateer raids contributed to the construction of the Pacific Ocean as a geopolitical space, a process that Magellan had started and that would come to an end with James Cook's voyages during the last quarter of the 18th century (Maroto Camino 2005).

As expectations toward the strategic triangle in the south of the continent waned and other plans contemplating the shores on North America were brewing, the English adventure beyond the equator caused the Spanish increase of their control over southern frontiers (Andrews 1981). Following Drake's incursion into the Pacific Ocean and the sackings of Valparaíso and El Callao ports in 1579, the Council of the Indies approved in 1581 an ambitious project that aimed at establishing two fortresses at the mouth of the Strait. The outcome of the enterprise led by Sarmiento de Gamboa proved to be very expensive as it implied operating in an unknown geography.[23] In addition, the first data on the insular nature of Tierra del Fuego spread and it discouraged the idea of maintaining military garrisons in the continent's inhospitable confines.

In fact, during the last decades of the 16th century, the first maps depicting a small archipelago at the end of the known mainland (America), separated from the unknown southern land by a new maritime passage started to circulate. The Dutch cartographical discourse foresees big commercial companies' interventions by considering the possibility of free access to the Pacific Ocean through this new route, as it is shown on 1590 Jodocus Hondius' map.[24]

The Netherlands took over the nautical leadership of Iberian Empires and, counting on the experience in the creation of trading companies to organize expeditions, set out to start trading in the South Pacific. Within the rivalry context with the Spanish Crown, the first goal was to discover trans-Pacific routes that linked American shores to southeastern Asia (Crespo Solana 2013).

B. Doetechum's map of the Magellan Strait (1618), made with Olivier van Noort's expedition observations, outlines the scope of the first Dutch voyages to the southern coasts. It also territorializes the shores of the interoceanic channel through toponomy and highlights—by means of a map scale change—the ports of call in the Atlantic and the South Pacific oceans: Río de la Plata, Puerto Deseado and Mocha Island. The iconography reveals two elements of interest. On the one hand, the vessels show the east–west fleets direction, indicating the trading route toward the South Sea through the Strait and up to the cities in the center of Chile considered as potential markets. On the other hand, the huge penguin population typical of southern Patagonia represents the possibility of getting an important resource to feed the crews (Fig. 3.5).[25]

After three expeditions that crossed the Magellan Strait—Jacobo Mahu and Simón de Cordes (1598–1599), Oliver van Noort (1599–1600) and Joris van Spilbergen (1614–1615)—Netherlands States-General granted exclusivity over that route and of the traditional African route to the Dutch East India Company (VOC).[26] A former promoter of this entity, Isaac Le Maire, founded a new company in order to discover another more southerly route and establish supporting bases in a marginal territory subject to Spanish imperial domain. The expedition led by William Schouten and Jacob Le Maire (1615–1616) complied with the first goal, upon confirming Santa Cruz' hypothesis and Drake's information on the insular nature of Tierra del Fuego,

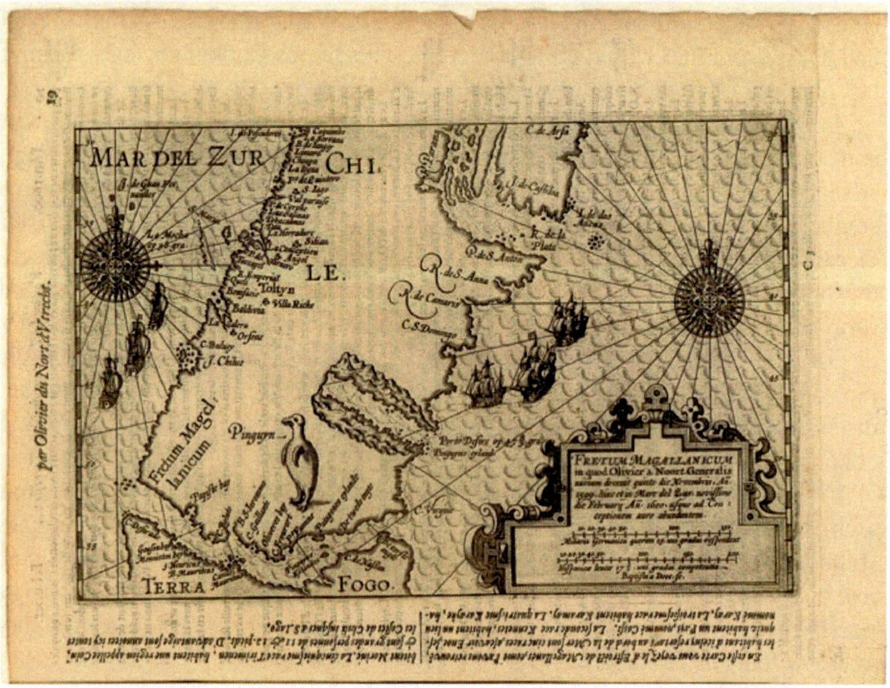

Fig. 3.5 van Noort *Fretum Magallanicum* (1618)

by corroborating the existence of a second passage to the Pacific Ocean through Cape *Hoorn* (Schouten 2010).

Spain's response was to organize a new exploratory voyage led by the Bartolomé brothers and Gonzalo García Nodal, aimed at circumnavigating the archipelago and verifying these findings. The *Carta de Tierra del Fuego con los Estrechos de Magallanes y Le Maire*[27] produced in 1618 for this expedition by Portuguese cartographer Pedro de Lettre was based on Le Maire's map published that same year. It is an antecedent of Hondius' map *Freti Magellanici ac novi freti vulgo Le Maire* (ca. 1620), a model of the representation of the Magellan territory during the 17th and the 18th centuries (Fig. 3.6).[28]

Pedro Texeira Ealbernas elaborated a map that accompanied Nodal's account, published in 1621, whereas cosmographer and chief of navigation Diego Ramírez de Arellano drew another chart with the data collected in the exploration from 46° to 59° south latitude.[29] The latter challenged the authorship of the first map, which showed the actual geographical entity of Tierra del Fuego. The Council of the Indies resolved the issue, by commissioning the merging of both maps. Cartography vindicated the insular edge as the Spanish Empire frontier, by replacing Dutch and English toponyms with other Spanish names, identifying the southernmost tip at the small rocky archipelago called *Diego Ramírez* and renaming Cape Horn as *San Ildefonso.*[30]

Following the expedition of the brothers Gonzalo and Bartolomé García de Nodal, there was a debate on the role, risks and usefulness of the Straits of Magellan and

Fig. 3.6 Hondius Freti Magellanici ac novi freti vulgo Le Maire (ca. 1620)

Le Maire and the new southernmost passage, in the framework of the defense of the Pacific shore in the Western Indies, the Philippines and other alleged southern lands. Due to the failure of Sarmiento de Gamboa's project, the fortification of the region was ruled out. Instead, the idea was to use the region as a communication, provision and operations of a war fleet area. The re-foundation of Valdivia was also an alternative, though it was not eventually an immediate imperial priority (Bradley 1979; Noejovich and Salles 2011).

Spain was content with a symbolic appropriation of the territory, relying on the fact that harsh weather conditions would deter its rivals. Conversely, the Dutch started to use the southern passages when navigating to the East, with ports of call in Brazil and Patagonia. Some maps produced during the first half of the 17th century reflect the expectation of building a support area in the Pacific Ocean, by founding enclaves or trading posts (factories) on Chilean shores, and also, by seeking alliances with the natives. J. L'Hermite's expedition (1623–1626) and that of H. Brouwer (1642–1643), both organized after the 12-year truce between Spain and the United Provinces (1609–1621), intended to establish a Dutch settlement in southern Patagonia. Joris van Spilbergen's description of the weakness of the Spanish defense in the American Pacific Ocean and of the fertile lands of Chile—rich in metals as well—together with the possibility of getting help from native inhabitants that had resisted the conquest fueled the project (Schmidt 1999). Although the attempts of occupation failed, due

Fig. 3.7 Gerritsz *Mar del Sur. Mar Pacífico* (1622)

mainly to the lack of support of the chieftains when urged to locate gold mines, an intense exploratory labor began, which contributed to a cartographic differentiation of a territory that so far had been neglected, since attention was focused solely on the Magellan Strait.

Hessel Gerritsz and Frederik de Witt's maps account for the intention of Dutch companies and the advance of a conception of a bi-oceanic southern region that, in geostrategic terms, was paramount for controlling spices trade. Gerritsz, chief cartographer of the Dutch East India Company, produced the map *Mar del Sur. Mar Pacífico* (1622), with a frame quite similar to Ortelius' (1589), differentiating with these toponyms the ocean mass to the north from the ocean mass to the south of the Equator. He plotted the route followed by Schouten and Le Maire's expedition, outlining an itinerary that attempted to avoid a dispute with Spain. Sailing from Puerto Deseado along the eastern coast of Tierra del Fuego and, upon confirming there were no lands up to 60° south latitude, they headed to the Pacific and entered into Juan Fernández and Mocha Islands. Finally, they adopted the usual course in previous circumnavigations to the north of the Tropic of Capricorn, setting course for the Moluccas, Solomon and New Guinea islands. The fleet's depiction shows the different routes and navigability conditions and warns of the perils of such an enterprise along southern latitudes by representing ships struggling in a storm. In the South Atlantic Ocean, he located an archipelago of three islands, which he named after Sebald de Weert, who had spotted it in 1599, and included two channels across

Fig. 3.8 Gerritsz *Provincien vande Straet van Magallanes, ende vande Straet Le Maire* (1630)

the Isla Grande of Tierra del Fuego, raising expectations of finding alternative routes less risky than the Magellan route. Dotted lines indicate both channels, showing lack of enough evidence. The channel appearing more to the north is identified as *Entrada de San Sebastián*, while the one to the south—nameless—could indicate, as Martinic (1999) remarks, the sight of the eastern entrance of the channel, which was identified two centuries later by the British and named Beagle Channel.[31] Later Dutch maps reproduced the entrances of both channels, although due to the prevalence of the configuration posed by the *Tabula Magellanica* de Blaeu (1640), they kept the hypothesis of the San Sebastián channel (Fig. 3.7).

Gerritsz' map, published by Johannes De Laet in 1630, could be considered—regarding the spatial frame and the selection of coastal toponyms—the depiction of a bi-oceanic region between 37° and 60° south latitude. It identifies key sites for navigation and for the arrival of fleets: the Malvinas Islands, the Magellan Strait, Cape Hoorn and the Le Maire Strait, and the *Staaten landt*, represented with open coasts to the east (Fig. 3.8).[32]

F. de Witt's map (1675) presents an approach similar to the 1630 Gerritsz map, since it denotes the ongoing interest in the southern region and vindicates former maritime enterprises in a time of Dutch naval power decline. The illustration at the top manifests the intention of an Alliance or commercial exchange with the Mapuches and the willingness to obtain gold. The representation of a naval battle in the South

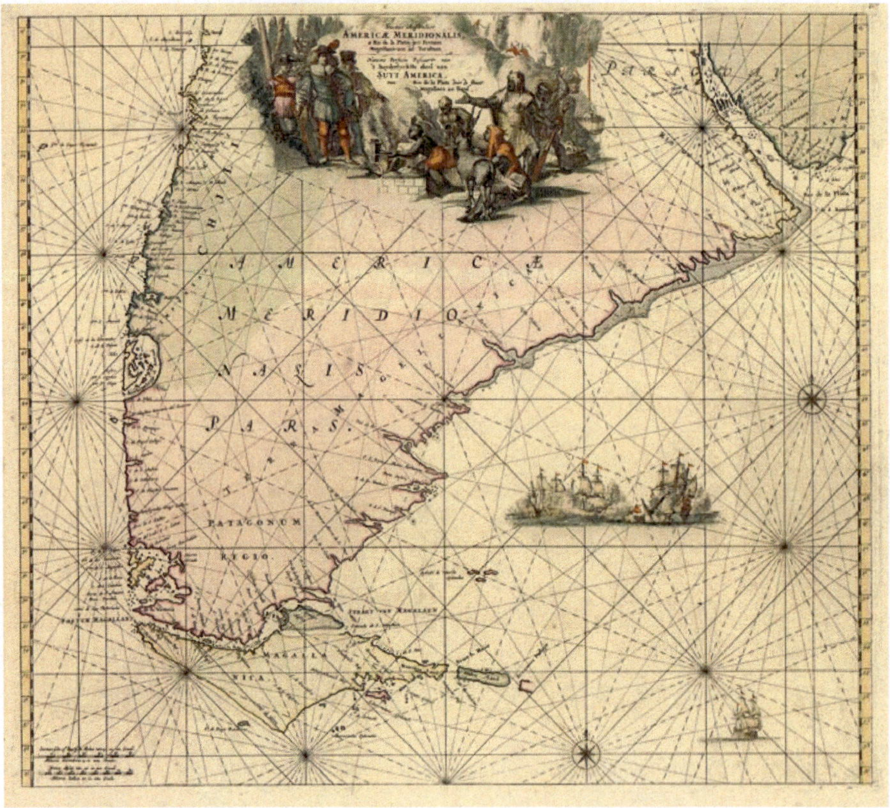

Fig. 3.9 F. de Witt *Tractus australior Americæ Meridionalis* (1675)

Atlantic Ocean manifests the existing rivalry for the control of the maritime space where the Sebald de Weert islands were strategically located. It portrays a passage or strait—Brouwers—between *Isla de los Estados* and *Terra Australis*. Conversely, a previous version produced by Donckert in 1659 omits it, definitely severing the archipelago of Tierra del Fuego from the alleged southern lands (Fig. 3.9).[33]

As of the mid-17th century, the southernmost tip of America started to gain its own identity within European cartography and, concurrently with the new maritime and commercial expansion plans of France and England, maps show the differentiation of a bi-oceanic territory.

Nicolas Sanson's map (1657) identifies the region with three main toponyms: *Mar Magallánico*—South Atlantic and South Pacific oceans—*Tierras Magallánicas*—Patagonia—and *Islas Magallánicas*—archipelago of Tierra del Fuego. It incorporates plenty of toponyms on the coasts with names of Hispanic and Dutch origin and recovers the memory of the first settlements in the Strait of the 16th century. The delimitation of the Kingdom of Chile at the left upper corner also contributes to differentiating the Magellan region from the northern regions of Chile and Tucumán together with their adjoining seas (Fig. 3.10).[34]

Fig. 3.10 Sanson *Detroit de Magellan, Terre et les Isles Magellanicques* (1657)

The charts elaborated on the information provided by John Narborough's voyage of exploration (1670 and 1671) to the Magellan Strait also contribute to the cartographical construction of a region, in geopolitical terms.[35] The outcome of the expedition reflects not only the advance in the hydrographical knowledge of the interoceanic channel if compared to that of the Dutch, but also the interest in the nature of the territory and its inhabitants. The nature of the information contained in the voyage cartography and account, published not before 1694, as well as the taking of possession that took place on the coasts of the Strait and Puerto Deseado, evince the two-fold purpose—scientific and political—and the intent of assessing the possibilities of a future territorial occupation. Thus, cartography anticipated the space of dominion, updating the strategic triangle thought in the first expansion plan that had originated in Drake's voyage. John Thornton's map (1674) enables a political reading, since it announces in its title "Discovery" of the Strait by Narborough, and manifests the intent of territorializing through toponymy, by calling the southern sector of Tierra del Fuego as *King Charles South Land.* Likewise, it provides a simplified version of the abundant hydrographical information produced by the expedition—used as a standard navigation guide of the Magellan Strait for decades—and "fills" the Patagonian interior with a frame that outlines the area of interest from the 40° south latitude to Cape Horn (Fig. 3.11).[36]

The manuscript map produced by pilot Francisco de Seixas y Lovera to accompany his *Descripción Geographica y Derrotero de la Región Austral Magallánica* (1690)[37] focuses its attention on a space, neglected by the Crown since the failure of Sarmiento de Gamboa's enterprise (1578–1584). It also shows the possibility of

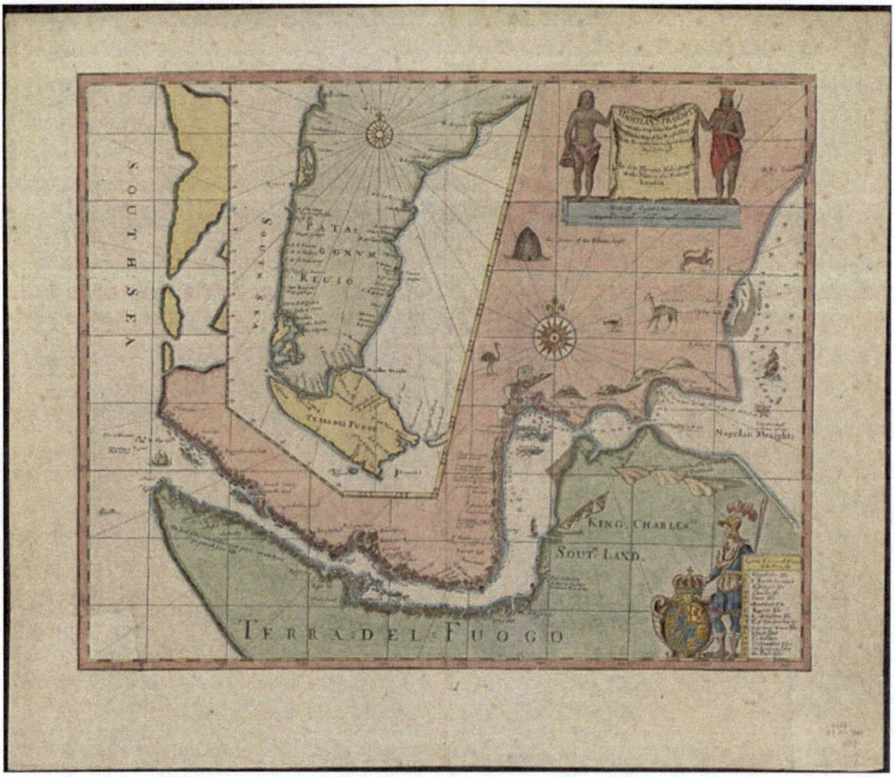

Fig. 3.11 Thornton *Magellan Straits* (1674)

foreign interventions by representing places in Tierra del Fuego, presumably inhabited and fortified by the English, since the mid-century, and in the north of the western mouth of the Strait, since 1675. Although the maritime reconnaissance performed from Chiloé was not able to confirm the metropolitan suspicion of an English occupation following Narborough's expedition, the cartographic story re-creates news that circulated during the 1970 s and 1980 s, generated in local testimonies that spread between 1674 and 1675. While a chono native called Talcapillán affirmed there were colonies on the coasts of the Pacific Ocean, between the Gulf of Penas and the Strait mouth, the information obtained from the natives who had accompanied Jesuit Nicolás Mascardi in his last voyage from the mission of Lake Nahuel Huapi stated that there was a settlement of white people in the Patagonian interior. As regards the English population in the insular tip, a possible source is a Jesuit map of South America, dated *circa* 1682,[38] that locates it in the southeast of Tierra del Fuego, the same as Seixas y Lovera (Martinic 1997; Urbina C. 2016) (Fig. 3.12).

Cartography produced on the data obtained from the first French voyages to the south of America shows the difficulties imposed by the use of southern passages did not hinder the trafficking of traders and privateers and the organization of the first state-supported companies. In fact, the interest in the South Sea never decreased, be it

Fig. 3.12 Seixas y Lovera *Región Austral Magallánica* (1692)

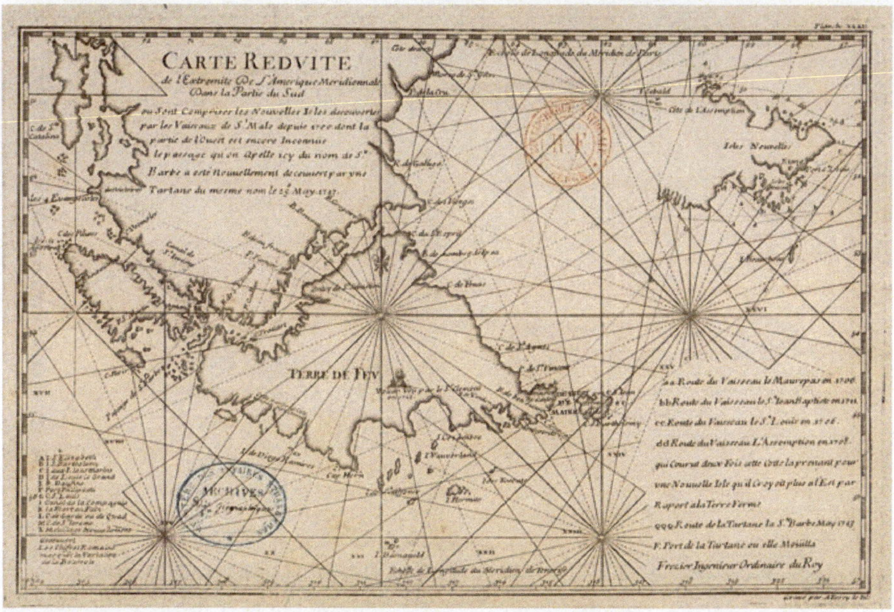

Fig. 3.13 Frezier *Detroit do Magellan et malouines* (1716)

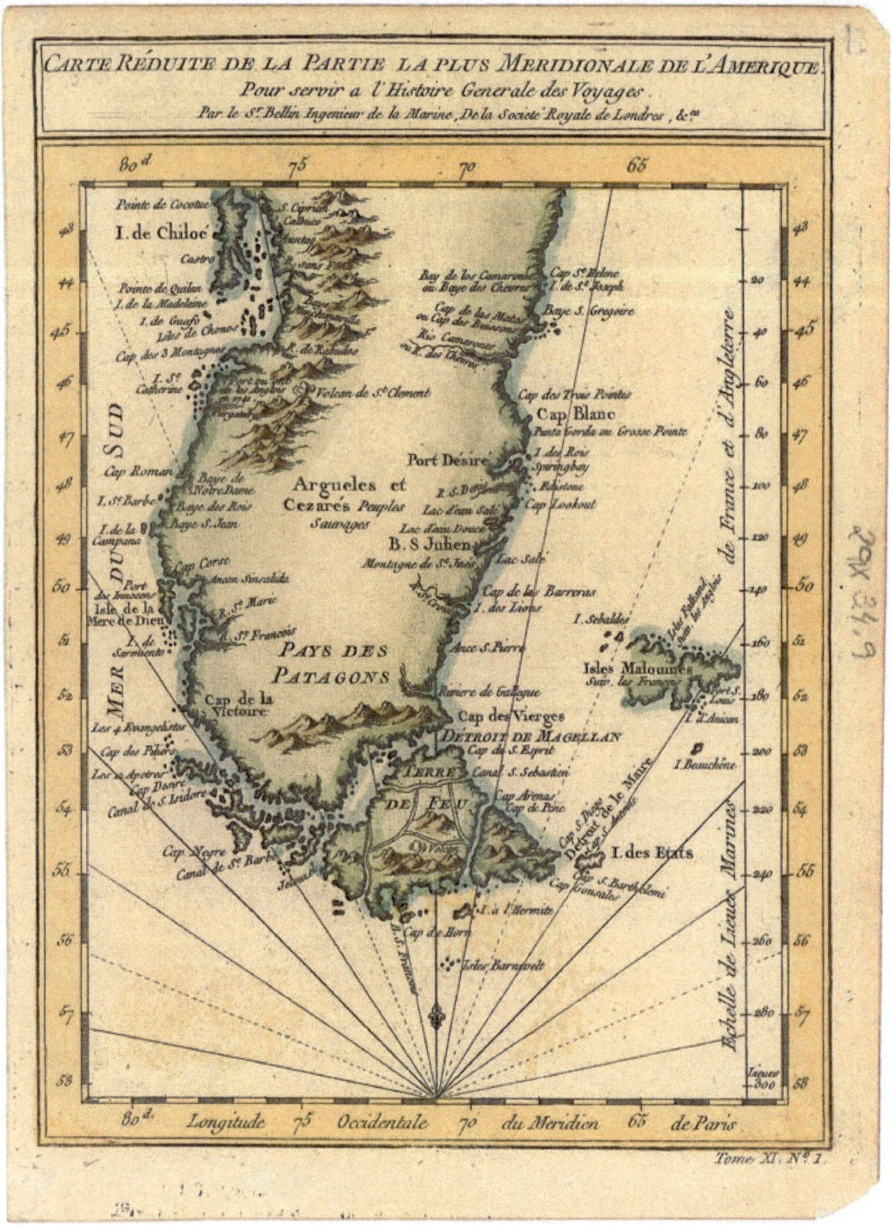

Fig. 3.14 Bellin *Carte réduite de la partie la plus meridionale de l'Amerique* (1740)

in terms of precious metals, commercial perspectives, conquest of lands or settlement of bases or enclaves (Bradley 1989). The expedition led by J. B. de Gennes (1695–1697) was just an incursion in the southern coasts of Chile. The next one, organized by the new South Sea Company, set out to broaden the knowledge of the Magellan region, in view of the settlement of a base or trading post, a project that was encouraged by the maps and reports of buccaneers operating on the American coasts of the Pacific Ocean. The expedition led by J. Beauchesne-Gouin (1698–1701) met the exploratory target and obtained good benefits from its transactions in Chile and Peru.

During the following years, French commerce expanded to China and the Eastern Indies. According to estimates, more than 150 French ships sailed along the South Pacific Ocean using the route round Cape Horn between 1698 and 1725 (Spate 1983). The increase of commercial trade was concomitant to significant map production, some of which had several re-editions and reproductions during the second half of the 18th century.

Guillaume de L'Isle's *Carte du Paraguay, du Chili, du Detroit de Magellan* (1703), by focusing the representation of *Terre Magellanique* and highlighting its coasts, offers the image of a bi-oceanic region with no other utility than that of a port of call. Having early sources (Vespucci, Magallanes and Sarmiento de Gamboa) as well as those of the 17th century (Brouwer, Narborough, de la Roche, Beauchesne-Gouin, Sharp and Halley), it provides relevant contributions to cartography and southern region toponymy. It also manifests its territorial nature through illustrations that accompany the lateral cartouche—a couple of Tehuelches and another one of canoers in their coastal habitat, which represent the stereotyped image of Patagonia inhabitants as barbarian and savage.[39]

The Magellan Strait and the Malvinas Islands chart that was produced by the French military engineer Amédée F. Frezier and published in 1716 with the account of his last voyage to the coasts of Chile and Peru while serving Spain (1712–1714) depicts from 50° to 57° south latitude. The map shows numerous French ships, most of them from Saint-Malo, to which Frezier attributes the discovery in 1700. The archipelago is named Nouvelles and the toponym Sebald Islands is given to three small islands to the west of the big island. He identifies Port St. Louis as most frequented port by different expeditions, by sketching their courses and referencing them. He ignores the advances achieved in Dutch and English charts and only represents the coasts known to the *malouines*, leaving aside the western coasts and those of the interior of the channel between the biggest islands.[40] A similar configuration with a layout of incomplete coastlines appears again on the map produced in 1740 by the engineer of the French Navy, J. N. Bellin, in which the islands present a double toponym, "Malouines" as the main one and "Falkland" for the English.[41] Taking into account that there were no significant changes as regards the state of knowledge on the coasts of Patagonia and Tierra del Fuego when compared to the first half of the 17th century, we note that the exploratory interest focuses almost exclusively on the South Atlantic and the Malvinas Islands (Figs. 3.13 and 3.14).

3.4 The Southern Region as a Dominion Space

The 18th century sources show the growing geopolitical and economic valorization of the South Atlantic Ocean, the importance of the Río de la Plata ports—Buenos Aires and Montevideo—and the Patagonian maritime front as a defense of the southern borders of the empire. The intensification of the use of the interoceanic passage through Cape Horn by the French and the English led the Spanish Crown to officially open this route, concurrently with bringing to an end the system of fleets and galleons around 1740 and adding hierarchy to the southern territories within the metropolitan strategy (Nocetti and Mir 1997).

The importance of the South Atlantic Ocean space in the imperial agendas is clear, considering the advances made as regards geographical knowledge and the cartographic representation of the islands identified as Sebald de Weert or Sebaldines, Nouvelles, Falkland, Malounies and Malvinas, a strategic enclave for the Pacific Ocean markets and operational base for whalers and sealers. Toward the mid-century, they sketched an approximate configuration of the archipelago with the representation of the whole outline (D'Anville 1748)[42] and, during the following decades—mainly with the settlement of French, English and Hispanic-"criollos"—they produced the first local maps (Ruiz Puente 1769,[43] Bellin 1771[44] and Pernetty 1771).[45]

The publishing of the work *Histoire des navigations aux terres australes de Charles De Brosses* (1756) revitalized the purpose of discovering and annexing new lands in the southern hemisphere, present in the French imperial imaginary since the 16th century; particularly after the Seven Years' War (1756–1763), in order to compensate the loss of territories in Canada. With the expectation of "discovering those vast regions beyond the southernmost tip of the known world", the Malvinas Islands were to serve as a base for seeking the expected "fifth part" in the South Atlantic Ocean. The editor of the second edition of "*Historia de un viaje a las Islas Malvinas*" ("History of a trip to the Malvinas Islands"; 1770), written by the chaplain of the expedition led by Louis-Antoine de Bougainville, clearly stated: "If there has ever been a favorable moment to reach the southern world, it was right now, when Europe is convinced of the existence and position of the Malvinas Islands" (Pernetty 2012).

The expedition, sponsored and commanded by Bougainville (1763–1765), took possession of the islands in the name of the King of France and set up a base on the eastern island with seamen mostly from Saint-Malo. The colony numbered over 100 people and survived just for 3 years (1764–1767), since due to the Family Compact between the Bourbon Kings, Spanish sovereignty on the islands was recognized and the cession of Port Louis was decided, following a compensation to Bougainville and to the French government itself (Bougainville 2005; Pernetty 2012).

The English strategy also contemplated the occupation of the Falkland since George Arson's recommendations in his work *A Voyage Round the World* (1748), after his circumnavigation voyage in 1740–1741 (Goebel 1982). The secret instructions elaborated by the Admiralty for the expeditions of J. Byron (1764–1766), S.

Wallis and P. Carteret (1766–1769) and J. Cook (1768–1771, 1772–1775 and 1776–1779) evince the importance assigned to the islands as the "key to the Pacific Ocean" and the ongoing search for the *Terra Australis* (Luiz and Schillat 1997; Frost and Williams 1997; Torres Santo 2004).

The expedition led by John Byron took possession of the archipelago in the name of the King in 1765 and both, Wallis and Carteret's next voyage and Cook's second voyage, contributed to the settlement of a garrison on Saunders Island (Trinidad Island). A fleet organized from the Río de la Plata forced the English to abandon the island in 1770, but due to later negotiations, Spain agreed to abandon Port Egmont. In 1774, England decided to leave, considering it an inconvenient site with very costly maintenance. However, England did not waive its rights and left possession symbols (Rose 1929; Wallis 1978).

The Malvinas Islands Chart (1771) by engineer J. Bellin, produced with information provided by the memoirs of French and English voyages, summarizes the geopolitical situation of the archipelago. It uses a double toponymy—*Isles Malouines ou Nouvelles*/*Isles de Falkland*—and emphasizes their strategic value by locating them in relation to the Magellan Strait. It also identifies the inhabited sites, differentiating the nature of each settlement—"French settlement of Saint Louis/lodge of the English at Port d'Egmont" (Fig. 3.15).[46]

During his second voyage, following the instructions of circumnavigating the southern hemisphere through extreme latitudes, Cook crossed the Antarctic Circle

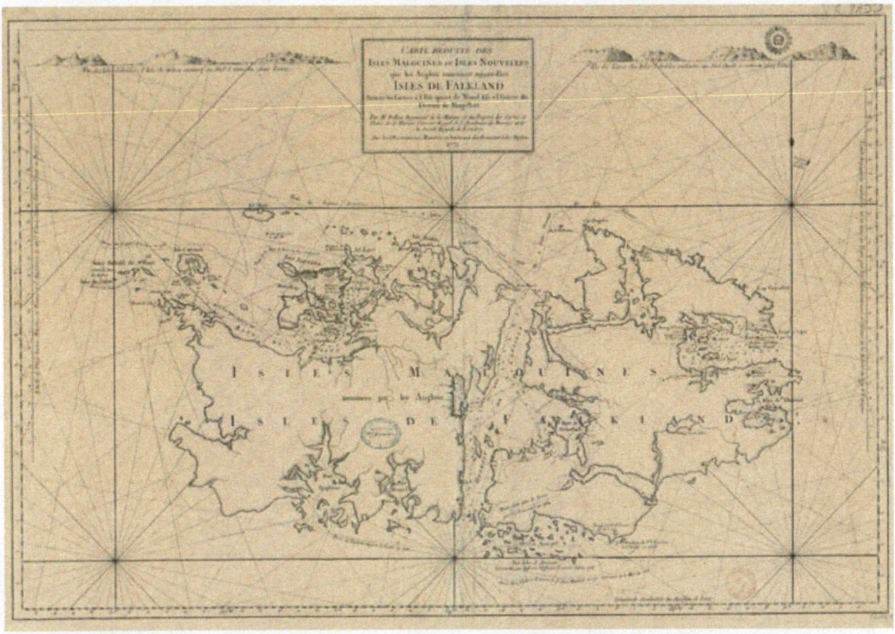

Fig. 3.15 Bellin *Carte reduite des Isles Malouines ou Isles Nouvelles* (1771)

several times and in January 1774 reached the 71° 10′ south latitude. The corroboration of the absence of a temperate southern continent in the South Pacific Ocean and the presence of icy lands to the south of the 60° confirmed the utility of a base in the South Atlantic Ocean. This would be part of a strategic global plan whose objective was to strengthen the British Empire through the control of maritime routes (Luiz and Schillat 1997). The chart of the southern extremity of America (1775) outlines the space of interest by representing a South Atlantic Ocean region between 47° and 57° south latitude, which comprise southern Patagonia, Tierra del Fuego, the Malvinas Islands and sub-Antarctic island (Fig. 3.16).[47]

Although England temporarily abandoned its plan of having a permanent settlement in the South Atlantic Ocean, cartography shows the ongoing explorations and the validity of territorial claims within the context of growing competition for the use of faunal resources. While the Royal Navy provided technical assistance as regards sailing,[48] the maps of South America of Jefferys (1776),[49] D'Anville (1794),[50] Wilkinson (1806)[51] and Cary (1807)[52] evince the prevalent representation of the Malvinas Islands as a British dominion or interest area. As Harley remarked (2005) in his essay on the geographical atlases produced in England during the 18th century, the State subtly kept the monopoly of cartographical power, by conferring private

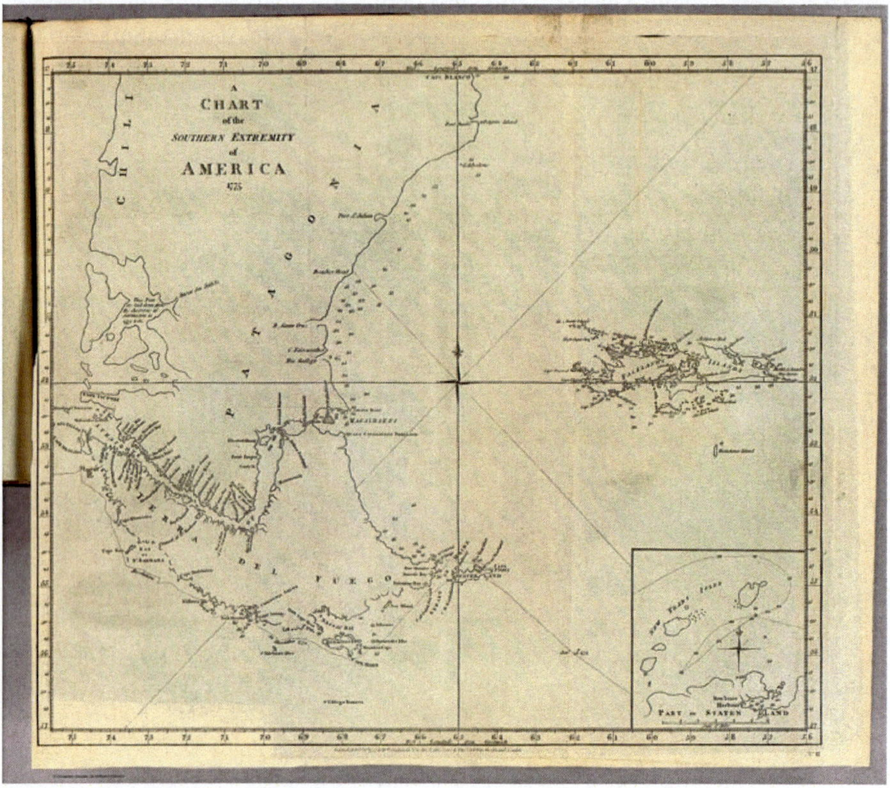

Fig. 3.16 Cook *A chart of the Southern extremity of America* (1777)

cartographer's freedom to determine the content and design of the works. However, the readers' preferences and needs—nobles, aristocrats and clergymen, apart from the Crown officials—and, above all, the internal logic of printed cartography—that is, set codes and procedures that structured practice and representation[53]—ensured the reproduction of geographies built from a political approach.

Together with other strategies, repetition and standardization of data constituted an effective means of legitimation of colonial interests. Maps that identify the Crown's overseas possessions show the concentration of internal and external power in the cartographical representation. *South America* (1776) of Thomas Jefferys, geographer to the King and active colonial agent, indicates the discovery of Malvinas islands by John Hawkins in 1593 and expresses the English supremacy in the Pacific Ocean–Great South Sea, representing the routes used in Narborough, Halley, Anson, Byron, Wallis, Carteret and Cook's expeditions. Thomas Kitchin's map of the world (1799),[54] part of the cartographical corpus typical of English imperial propaganda, highlights with coloring the different sites of interest in the southern hemisphere: New Holland (Australia), New Zealand, Malvinas islands, San Pedro or Georgia Islands and Sandwich Land—Southern Thule. The maps of Robert Wilkinson (1806) and John Cary (1807) distinguish by means of colors the Falkland and South Georgia Islands from the vast territorial entity comprising Patagonia and Tierra del Fuego. Far from ambiguity, the former refers to these archipelagos as British Islands (Figs. 3.17 and 3.18).

The actual or potential aspirations of rival States and the presence of whalers and sealers of different nationalities that regularly visited the southern shores led

Fig. 3.17 Jefferys *South America* (1776)

Fig. 3.18 Kitchin *A new map of the Word* (1799)

the Spanish to consider the need of effective occupation and the implementation of control strategies in an area that was difficult to defend because of its vastness and remoteness. The attention paid to the coastlines of Patagonia, Tierra del Fuego and Malvinas Islands is manifest in the pondering of missionary projects that resulted in being functional to the geopolitical interests of the Crown (Navarro Floria 1994; García Insausti 2017). The first attempt carried out from Buenos Aires during the settlement of Jesuit reductions to the south of the Salado River failed due to the report of Captain Olivares y Centeno,[55] in charge of the coast reconnaissance performed in 1745. Also, the observations of missioners Quiroga, Ströbel and Cardiel (Lozano 1969), members of the expedition, listing the disadvantages of peopling the Patagonian shores, led to such failure. A few decades later, other two projects of members of the Order who combined evangelization tasks with the defense of Spanish sovereignty in the southern territories got to the Bourbon administration. Due to the shipwreck of the ship named *Purísima Concepción* at the southeast of the Isla Grande de Tierra del Fuego in 1765, the idea of settling a mission on that coast reappeared. Juan Nepomuceno Walter of the Province of Chile had planned it the year before in the evangelization project, for being useful as a base for those sailing through the Le Maire Strait. The second proposal, anonymous and possibly elaborated in 1766, accounted for the convenience of founding a colony on the Malvinas Islands with a three-fold purpose. Firstly, to undertake the evangelization of natives from Patagonia and Tierra del Fuego who were isolated from other inhabitants. Secondly, to assist sailing by providing a repair and provisioning port and, finally, to exercise a greater territorial control by means of a military garrison in the southern region (Ratto 1930; García Insausti 2017).

Although the expulsion of the Order in 1767 hindered the realization of these plans, that same year Spain started the occupation of the Malvinas Islands. After the French left Port Louis, the place—depending on the Captaincy-General of Buenos Aires—was the site of a colony and a prison until 1811 (Levaggi 1978). The chart of the islands ordered by the first Governor Felipe Ruiz Puente, after establishing himself in Puerto de la Soledad, is the result of lands and shores reconnaissance performed in 1769 in order to spot possible foreign settlements.[56]

***Plano de las Yslas Malvinas* (1769)**
http://bdh-rd.bne.es/viewer.vm?id=0000143557

The funds from the *Real Situado* devoted to the maintenance of the base at the Malvinas,[57] the launching of an enterprise fully commissioned and sponsored by the State in order to start peopling Patagonia[58] and the costly maintenance of the forts[59] evince the importance of the South Atlantic Ocean in the Bourbon policy. Although they focused on political-strategic factors, they also pondered the economic advantages of exploiting marine resources and salt flats, as well as the possibility of withstanding incursions of the native inhabitants to favor the development of cattle breeding of Buenos Aires. Concurrently, the decrease in border conflicts and the chance of counting on the natives' cooperation contributed to lessen the risks of an eventual alliance with foreigners (León Solís 1994; Luiz 2006).

Cartography shows the valorization changes as regards the southern region, though it still re-created images of the continental interior as a vast unknown space controlled by independent nations. This accounted for the limited scope of the colonization policy. Cano y Olmedilla (1775) and Sáa y Faría's (1786) maps enabled an intellectual and symbolic appropriation of space. However, upon distinguishing colonial and indigenous jurisdictions by means of several cartographical resources—coloring, toponyms and ethnonyms, legends and other inscriptions—they acknowledged native control over southern territories.

The *Mapa Geográfico de América Meridional* developed by Juan de la Cruz Cano y Olmedilla[60] between 1765 and 1775 with documents from the Council of the Indies and an extensive corpus of maps and charts, including those produced by the Jesuits, shows an image prior to the creation of the Viceroyalty of the Río de la Plata (1776). It is very likely that they planned the occupation of the Patagonian shores on this production, which is one of the most important in the cartography produced during the period of Bourbon reforms. It represents the territories to the south of the Negro River as a space of dominion, by using the following toponyms: "*Kingdom of Chile*" and "*Modern Chile*", together with a legend that gathers valid names in the imaginary of the region since the 16th century: *Tierra Magallánica, de los Patagones y los Césares*. It describes its territorial nature by identifying the groups with whom they kept frontier relationships: "*Aucas, Puelches, Toelchús and Serranos from whom other partialities are derived and who are in contact with the Spanish.*" It also represents the southern frontier of the territories actually inhabited, by integrating natives who took part in the cattle trading circuit and who had commercial and diplomatic exchanges. In the "*Comarca Desierta de Patagonia*", the toponymy enables the construction of a colonial history by locating the establishments "*Nombre de Jesús arruinado*" and "*R. San Felipe*", first antecedent of a settlement that confirms Spanish sovereignty since the 16th century. In the Magellan Strait, it identifies some coastal features with native toponymy used in his Maritime Map (1769).[61] It differentiates Patagonia and Tierra del Fuego from the Malvinas or the Falkland Islands by means of colors and locates on said islands the settlement sites, Egmont or de la Cruzada and B. de la Anunciación.

The *Mapa Geográfico* (1786) developed by the Brigadier José Custodio de Sáa y Faría,[62] commissioner of limits serving the Portuguese Crown since 1750 and the authorities in Buenos Aires since 1777, evince important advances in the configuration of the Atlantic coastline, as a result of the European exploratory work, and in the knowledge of the Patagonian interior. This map represents the basin of the Negro and Santa Cruz Rivers as interoceanic communication axes, using the information provided by the reconnaissance carried out with the guide of native inhabitants, and identifies the unexplored sectors with the legend "Land inhabited by various Nations of barbarian natives". It indicates that the extension and unfamiliarity of the land and the fact that several independent groups occupied it hinder its penetration. This map can be considered a synthesis of the geographical knowledge of the time. However, the omission of relevant data for the territorial dynamics understanding, which were available in the documentary corpus checked by the cartographer, let us draw the conclusion that the selection of information was conditioned by the political goals of its promoters within the debate on the useful(less)ness of Patagonia.

Sáa y Farias' reports, elaborated by request of Viceroys Sobremonte and Loreto, contain statements that accounted for the option of a restricted settlement, limited to strategic sites of the Atlantic Ocean front. In the document that examines the results of Villarino's reconnaissance, dated on 24 October 1783, the cartographer stated, "It is true that the Negro River does not provide any advantages, neither for commerce—due to the poor land—nor for agriculture—due to its sailing difficulties". Later, he added, "even acknowledging the navigability of said river, I consider it unfeasible, having its course for a few unknown and uninhabited fields" (Sáa y Faría 1972: 1149). The report that contains the *Mapa Geográfico*, dated on 2 September 1786, also presented three precise recommendations. In the first place, it recommends peopling San José, considering its suitable conditions of the port for the ships to stopover and the possibility of making use of cetaceans and salt flats. Secondly, it recommends keeping the fort of the Negro River as a center of provisioning of cattle through trading with natives and finally, founding another fort in the site where the Jesuit reduction Nuestra Señora del Pilar had existed, aiming at facilitating land communication among the settlements in Patagonia and Buenos Aires.[63]

In the end, Sáa y Faría elaborated an argument that supported the dominion over southern territories in the Río de la Plata up to the Magellan Strait and the Malvinas Islands and, concurrently, contributed to account for the limited scope of the settlement representing Patagonia as a vast unknown territory controlled by indigenous people.

From the metropolitan perspective, shared by some high officers of the Viceroyalty, the occupation of strategic sites on the southern shores was enough to deter any foreign attempt to do the same. Likewise, it was sufficient to exert control over interoceanic navigation routes and to keep maritime activities of rival powers under close surveillance. Sáa y Faría's *Mapa Geográfico*—as well as the cartography that resulted from the surveys carried out by Antonio de Córdoba (1785–1786 and 1788–1789) and Alejandro Malaspina (1789–1795)—reinforce this approach by representing the maritime coastlines as an appropriate space through toponymy.

The *Carta esférica* (1798) engraved under the direction of J. Espinosa y Tello and F. Bauzá y Cañás[64] presents a synthesis of the operative and cartographic knowledge of the southern region by the end of the 18th century. Elaborated on the information collected in situ by the expedition led by Alejandro Malaspina (1789–1794),[65] it differed from Cano y Olmedilla and Sáa y Faría's maps, produced in a cabinet. Since one of the major goals of the voyages was surveying the coasts and outlining maps and courses, the officers in charge of the hydrographic work performed an extensive work, systematizing previous knowledge and contrasting, correcting and broadening information with their own observations, carried out with the most modern equipment of the time (Luiz and Gullón Abao 2014). They gathered all this in an important corpus of documents.[66]

The coastlines with more concentration of toponyms indicate the most relevant sites in the itinerary and the places with more strategic value—Malvinas and Chiloé archipelagos and the Strait of Magellan. Meanwhile, the coastal profiles, placed in the "blanks" that appear in the unexplored interior, identify key sites for the navigation between the two oceans at different latitudes (Cabo Blanco/San Jorge Gulf, Puerto Deseado, Madre de Dios Island, Savage/Malvinas Islands, Cape Horn, Diego Ramírez Islands and Mocha Island). Penhos (2005: 321 et seq.) refers to the complementary character of the cartographic and iconic records in the organization of visual data. The profiles of geographical features and the sights of the cities are representations of territory as from the ships; that is, partial and fragmentary sights of a space difficult to envisage in its vastness and variety.

The map of southern America also manifests the limits of the exploratory program illustrated in the representation of the archipelago of Tierra del Fuego, by depicting the alleged maritime passage that connected the Strait of Magellan to the Atlantic Ocean, leading into San Sebastián Bay. Bearing in mind that Cook (1775), an important source in the Malaspina expedition, does not represent this channel, we can assume that Bauzá found in the dotted outline as a way to pose the geographic enigma without discrediting other sources (Ovalle 1646; Cardiel 1751; Falkner 1772; Cano y Olmedilla 1776 and Córdoba 1786). This state of (lack of) knowledge of Tierra del Fuego lasted various decades, as shown in the instructions given by the Admiralty to R. Fitz Roy in November 1831. Among the goals of the exploration, they included "to foster dialogue over the dispute on San Sebastián Channel" (Fitz Roy 2013: 46).

The *Carta esférica* achieved a significant advance in the representation of the coastlines when compared to the maps of Cano y Olmedilla (1776) and D'Anville (1779). That is the reason why the French Marine and the cartographer W. Faden published it again. It amounted to an important source for future surveys and cartographic production until the mid-19th century (González Leiva 2004).

The map published by Faden broadens the scope of action by incorporating the Juan Fernández Island, the first port of call used by the Dutch and English seamen who sailed the Pacific Ocean through Cape Horn and the Río de la Plata estuary. The said estuary was considered one of the most important places in South America as

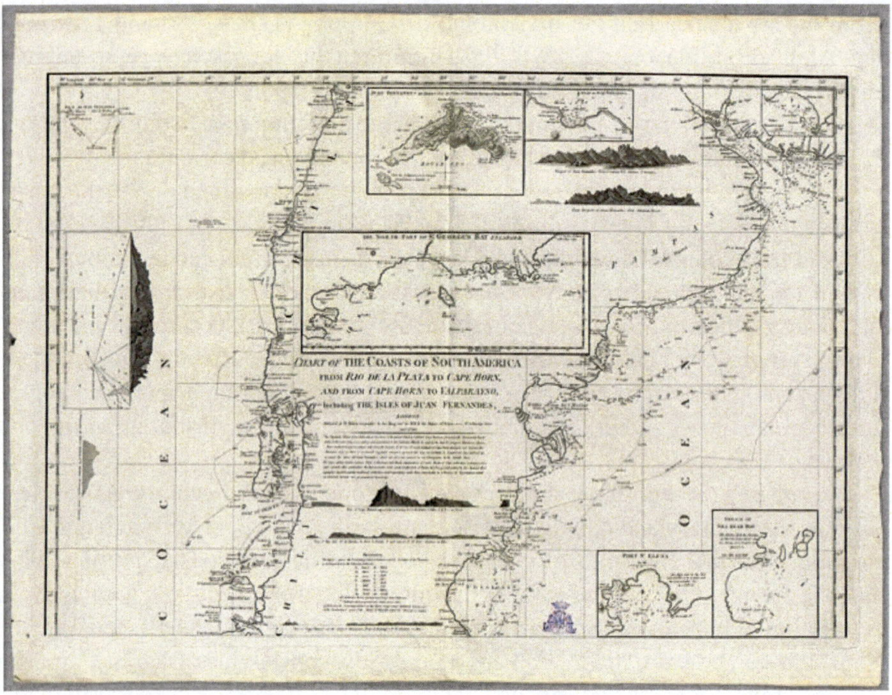

Fig. 3.19 Faden *Chart of the Coasts of South America* (1802)

regards commerce[67] due to its concentration of people and riches. While the spatial frame respecting the latitude and longitude of Bauza's chart provides the image of a bi-oceanic southern region, the depiction of the ports on the Patagonian and Chilean shores evinces that the interest kept focusing exclusively on the coastlines. In the interior, the toponym *Chica* insists on the idea of a land of uncertain utility and contributed to the devaluation of the territory, which lingered up to the 19th century (Fig. 3.19).[68]

The examination of the late colonial period cartography enables us to note that the Bourbon territorial reorganization program did not consider an accurate division of colonial administrative units in the southern territories in Chile and Río de la Plata. Therefore, it was difficult to apply the *utis posidetis* principle upon defining the limits of the new States during the 19th century (Luiz and Schillat 1997).

Cano y Olmedilla's map (1776) is a good example of the lack of concern as regards the internal jurisdictional limits. Although in a later edition the name "New Viceroyalty" appeared in the place of "Buenos Aires Province or Governorate", no changes took place on the territorial image. Thus, the Cuyo region remained within the Captaincy General of Chile. Considering the context within which it was elaborated, it is very likely that they prioritized the demarcation of the controversial border with Portugal to the east of the Río de la Plata and the Paraná River, disregarding the formal jurisdictions of the territories not yet incorporated to the effective dominion (Lacoste 2003). These are represented by differentiating two distinct entities through colors: Patagonia and Tierra del Fuego (green) and the Malvinas or the Falkland Islands (pink).

Miguel de Lastarria's production evinces the changes operated in the cartographic narrative and that of the documentation in general, as regards the jurisdictional conceptions. It describes regions or provinces using the best-known natural boundaries, as stated on the *Mapa Geográfico de América Meridional* (1804). This map shows a space subject to Spanish dominion up to Cape Horn, incorporating Tierra del Fuego to the Kingdom of Chile, and Patagonia and the Malvinas Islands to the Buenos Aires Viceroyalty. Instead, the *Carta Corográfica del Virreinato de las Provincias del Río de la Plata* (1804) represents just the effective jurisdiction up to 35° or 36° south latitude, and indicates the "current border" from the pampas to Cuyo Provinces and the "projected border" to the south of the Salado River. This chart identifies the indigenous territories by means of the change of color and a legend stating "gentile nations roam" to the south up to the Strait of Magellan and that the only Spanish settlement was "a small settlement by the Negro River on the Patagonian shore and the adjacent Malvinas Islands".[69] It represents the Andes as a natural and jurisdictional boundary between the Río de la Plata and the Kingdom of Chile—an image that was reinforced and still present in cartography since the 16th century—(Vega Palma 2006) and it insinuates that the Bío Bío River constitutes the southern border of the Kingdom of Chile.

Colonial administration kept elaborating frontier advancement plans that implied territorial expansion as well as the consolidation of land communication routes across the Andes,[70] and the state of knowledge enabled the identification of places suitable for settlements, routes and strategic sites. However, none of the projects came to reality during the colonial period. After pondering the costs and benefits, they noticed that they were carrying out the purpose of imperial policy without swaying away from the commercial circuit. They had already consolidated it and it was offering benefits to the markets with a direct connection on both sides of the Andes. Malaspina's conclusion over the boundaries of the empire issue affirmed the decision of a limited occupation, expressing the inconvenience of expanding the settlements beyond 40° to the south on the eastern slope of the Andes (latitude of Chiloé Island) and the mouth of the Negro River, on the western slope (Navarro Floria 1994).

The Patagonian settlement amounted to an ephemeral occupation of San Julián Bay (1781–1783), with a guard at the Port of San José (Valdés Peninsula). By the Negro River, it just constituted a few precarious and unfinished walls of the fort set in its mouth between 1779 and 1781.

Upon worrying about the control of foreign activities in the South Atlantic Ocean, colonial authorities devoted resources to support the garrison and prison on the Malvinas Islands (1767–1811) and granted occasional fishing licenses, sea mammals hunting and salt storing. Meanwhile, the metropolis promoted the installation of a factory ship at Deseado Port in order to affirm the Spanish presence in the region and to make use of the resources available for North-American, English, French and Dutch whalers' exploitation. The Royal Maritime Company, created by Royal Charter dated on 19 September 1789, operated with low revenues from 1790 to 1807.

On the Pacific Ocean side, the southern border of the Kingdom of Chile was defined at the Biobio River. To the south, the Spanish presence just involved Chiloé Island from the mid-16th century—the most ancient and permanent possession in the American South Pacific Ocean—the Fort Square of Valdivia since 1645 and the city of Osorno since 1796, spaces that depended upon resources provided by Santiago and extracted from the *Real Situado* (Lacoste 2003: 155-161).

Likewise, cartography as knowledge-power enabled the intellectual appropriation of space and contributed to the organization of spatial and territorial information for the design of borders policies. However, the indigenous response, based on a vast knowledge of the environment conditions that enabled accurate movement lines, routes control and efficient use of resources, still conditioned the Hispanic-"criollo" strategies.

3.5 Conclusions

The maps that expressed the discovery and exploration process in the southern hemisphere account for the different ways of producing knowledge through the interpretation of geographical data according to the needs and goals of the authors and/or promoters. The cartographical analysis enables the recognition of the imbricate scientific, territorial and economic interests in the overseas enterprises and visualizes how the objectives of the expansion defined the differentiation processes of space, territorial identities and geographic imaginaries.

The Atlantic Ocean discoveries during the first decades of the 16th century enabled the reappearance of the idea of a vast mass of southern land, as imagined since ancient times and considered essential in order to balance out the lands in both hemispheres. The idea was first proved on the southern shore of the interoceanic channel found in 1520 when Tierra del Fuego was considered by well-known Renaissance cartographers as part of the *Terra Australis* continent. Within the debate on the nature of those southern lands, the idea of an insular nature also had significant support. It was held by Santa Cruz, though with ambiguity in his *Islario,* and present as well in the production of some Portuguese and Italian cartographers, who ranged from island-archipelago to island of a big size (*isla de tierra*?), respectively.

The news on the insular nature of the lands to the south of the Magellan Strait, following Drake's itinerary up to the 56° or 57° south latitude, highly influenced the imperial strategies. On the one side, they discouraged the Spanish claim as to gaining control of the allegedly unique passage to the Pacific Ocean and, on the other hand, they prompted the organization of the first Dutch expeditions to the south of the continent.

The earliest antecedents of a geopolitical valorization of the southernmost tip appear in Spanish, English and Dutch projects to assure the use of the Strait and, since 1616, of Cape Horn, in order to access the American coasts in the South Eastern Pacific Ocean and the markets in the East. Cartography expresses such intentions, since the late 16th century and during the following century, by differentiating the

Magellan territory as a bi-oceanic region, strategic for the expansion to the Pacific Ocean.

During the 18th century, the need to broaden trading exchanges and to expand territorial dominions re-awakened the search for new lands in the southern hemisphere and the interest focused on the South Atlantic Ocean. Maps and journey accounts account for the growing importance given to the Malvinas Islands as a maritime station when sailing to the Pacific Ocean, a base of operations for the activity of whalers and sealers and the fulcrum of expeditions that intended to discover *Terra Australis.* Upon the occupation of the archipelago by France (Port Louis), England (Port Egmont) and Spain (Puerto Soledad), between 1764 and 1767, a South Atlantic Ocean region is shaped, anticipated in cartographical representations as a space of state interest. Maps of the end of the 18th century and beginning of the 19th century, with underlying motivations of a more aggressive imperialism, depict this space and assign them a specific geopolitical function.

Cartographical analysis also enables the recognition of some deep-rooted beliefs in the geographical imaginary of the southern region. On the one hand, the relationship between the tip of the island edge and the polar lands expressed initially by representing Tierra del Fuego as part of the southern continent and then in the conformation of an economic space that comprised hunting areas of sea mammals in the South Atlantic and Antarctic seas. On the other hand, we still observe the strategic value given to the maritime space and the lack of interest for the continental interior. Clear examples are the almost exclusive interest they showed in the reconnaissance of the shores and interoceanic passages and in representing geographical features and useful tips for sailing, ports of call and a base for hunting.

Cartography, consolidated as an imperial undertaking, contributed to the intellectual and symbolic appropriation of the southern space. It also participated in the construction of a geographical "reality" by posing representations in which both scientific and propaganda discourses fuse and confuse. In addition, even though they contained enough information to start understanding the functioning of the territory, maps continued showing spaces in white and other legends of the *terrae incognitae.* Based on Lois (2018: 15), the inaccessibility to uncertain geographies is not linked to the fact of not knowing them, ignorance or lack of information. It is rooted in the difficulty of creating representations that enable them been satisfactorily thought out, visualized and explained.

The same happened with other regions not yet explored, the southernmost territories in America fired the creative imagination, and journeys to the unknown, to the edge of the ecumene, never lost its charm. Travelers, from Columbus onwards, found in America what they were looking for and their testimonies show the tension between what they actually observed, what they knew and what they expected (Penhos 2005). The chronicler on Magellan's expedition stated his aim was "to confirm and tell about the wonderful things that could be seen when navigating the Ocean" (Pigafetta 1963: 41–42). Meanwhile, the editor of the second edition of Pernetty's work (1770) conceded: "We are so convinced that we are going to find amazing things that we get used to seeing them where they do not exist and sometimes to invent them"

(Pernetty 2012: 32).[71] In this period, cartographers contributed with their "inventions" to the wish of seeing. In the production process of "confirmed images" or "controlled inventions" (Harley 2005), they contrasted tradition, imagination, attraction for wonders and empirical knowledge, which started to gain ground with the broadening of geographical information and the legitimate status science grants observation as a warrant for reaching the truth.

Cartographic narrative shows the development of a southern hemisphere exploration agenda, fueled by scientific, territorial and economic imperatives, always intertwined. Upon communicating the illusion of broadening the "spatial panopticon" (Harley 2005), unveiling the boundaries, it uses its power on knowledge and strengthens the different States power while building new fields of action and legitimating future ambitions.

Notes

1. In current studies on territory, we approach territorial construction processes taking into account the interplay of legal, political, economic and cultural dimensions, considering practices and representations, strategies and the means used by different groups in order to appropriate certain functional or symbolic dominion over a geographic space share (Raffestin 2011; Haesbaert 2007; Haesbaert and Limonad 2007).
2. According to Padrón (2010), between 1507 Waldseemüller's map and 1529 Diego Ribero's map of the world, we observe a growth in the width of the Pacific Ocean that ranges from 80 to 134° longitude.
3. P. Reinel, *Carte du Monde*, 1519. Repository: National Library of France, Charts and Maps Department, Digital.
4. P. Reinel, chart of the southern hemisphere, 1524, reproduced in Destombes (1955).
5. O. Fine, *Nova et integra universo orbis descriptio,* 1531 Repository: Library of Congress Geography and Map Division-Washington, Digital. See Image in Chapter 2.
6. G. Mercator, *Planisfhere aux latitudes croissantes*, 1569. Repository: National Library of France, Charts and Maps Department, Digital. See Image in Chapter 2.
7. If the criterion of the distinction between continent and "*isla de tierra*" were the dimension of the landmass, the lack of accurate information on this regard in each case makes it impossible to determine when we are referring to one or the other one.
8. A. de Santa Cruz, *Islario general de todas las islas del mundo*, ca.1545. Repository: National Library of Spain, Digital.
9. A. de Santa Cruz, *Tierra o isla al mediodía del estrecho de Magallanes, Islario,* ca.1545. Repository: National Library of Spain, Digital.
10. In the 16th century geography books, the terms mainland, part and continent are synonyms (Lois 2018).

11. A. Thevet, *Les Isles de Sanson ou des Geantz – Le grand insulaire et pilotage* (1586). Repository: National Library of France, Charts and Maps Department, Digital.
12. J. López de Velazco, *Demarcación y navegaciones de Yndias,* 1575. Repository: John Carter Brown Library Map Collection.
13. B. Agnese, *Perú*, 1554. Repository: National Maritime Museum.
14. A. Pereira *North and South America*, 1545. Repository: John Carter Brown Library Map Collection
15. Homen Reinel, *Universa ac navigabilis to/tius terrarum orbis* [...], 1559, Repository: National Library of France, Charts and Maps Department, Digital.
16. D. Texeira, *Planisphere*, 1573. Repository: D. Texeira, *Planisphere*, 1573. Repository: National Library of France, Charts and Maps Department, Digital.
17. F. Vaz Dourado, *Atlas*, 1570. Repository: The Huntington Library, San Marino, California.
18. J. Martines, *Southern South America, Straits of Magellan with Tierra del Fuego,* 1578. Repository: The Huntington Library, San Marino, California. Image in Chapter 2
19. Among others, *Historia General de las Indias y del Nuevo Mundo* (1555) of F. López de Gomara (1555) and the *Cosmographia e geographia* of H. Girava (1555 and 1570). Repository: University of Seville Library, Digital.
20. B. Agnese, *Atlas* (manuscript), 1544. Repository: Hispanic Digital Library. Image in Chapter 2
21. G. Gastaldi, *Universale*, 1546. Repository: John Carter Brown Map Collection. *Carta Marina Nova tabula*, 1548. Repository: David Rumsey Historical Map Collection. Image in Chapter 2
22. F. Vaz Dourado, *Atlas*, 1570. Repository: The Huntington Library, San Marino, California.
23. Tomé Hernández' statement, one of the survivors rescued in 1587 by Thomas Cavendish, accounts for the tragic end of the colonists (cited in Barros 1978). The English expedition led by John Chidley and Andrew Merrick (1589–1590) rescued the only survivor of the Magellan colonies.
24. J. Hondius, *Americae Novissima Descriptio.* (1589). Repository: National Library of France, Charts and Maps Department, Digital. See Image in Chapter 2.
25. O. van Noort, *Fretum Magallanicum* 1618. Repository: John Carter Brown Library Map Collection.
26. The Verenigde Oostindische Compagnie (VOC) was since its foundation in 1602 until its dissolution in 1795 one of the most important and profitable trading companies operating in Asia (See Crespo Solana 2013, Keuning 1949)
27. P. de Lettre, *Carta de Tierra del Fuego con los Estrechos de Magallanes y Le Maire,* 1618. Repository: Royal Academy of History, Spain.
28. J. Hondius, *Freti Magellanici ac novi freti vulgo Le Maire,* ca. 1620. Repository: National Library of France, Charts and Maps Department, Digital.

29. Ramírez de Arellano, *Quarteroncillo fabricado del reconocimiento de los Estrechos de Magallanes y San Vicente,* 1619. Repository: National Library of Spain, Digital.
30. P. Texeira Ealbernas, *Reconocimiento de los Estrechos de Magallanes y San Vicen*te, 1621. Repository: National Library of Spain, Digital.
31. H. Gerritsz, *Mar del Sur. Mar Pacífico,* 1622. Repository: National Library of France, Charts and Maps Department, Digital.
32. H. Gerritsz, *Provincien van de Straet van Magallanes, ende vande Straet Le Maire,* 1630. Repository: Princeton University Library.
33. F. de Witt, *Tractus australior Americæ Meridionalis, a Rio de la Plata per Fretum Magellanicum ad Toraltum,* 1675. Repository: Princeton University Library.
34. N. Sanson, *Detroit de Magellan, Terre, et les Isles Magellanicques*, 1657. Repository: David Rumsey Map Collection, David Rumsey Map Center,, Stanford Libraries.
35. J. Narborough, *A map of the Straights of Magellan,* 1670. Repository: The British Library.
36. J. Thornton, J. Narbrough, & J. Seller, (1670) *A New Map of Magellan Straights Discovered by Capt. John Narbrough [...],* 1674. Repository: National Library of Australia Digital Collections.
37. F. de Seixas y Lovera, *Demostración de las Tierras y Yslas de la Región Austral Magallánica,* 1692. Repository: Library of Congress Geography and Map Division, Digital.
38. *Mapa de América del Sur*, ca. 1682. Archivo General de Indias, Mapas y Planos, Perú-Chile 176. Reference taken from Urbina C. (2016).
39. G. de L'Isle, *Carte du Paraguay, du Chili, du Detroit de Magellan*, 1703. Repository: David Rumsey Map Collection, David Rumsey Map Center, Stanford Libraries. Image in Chapter 4.
40. A. F. Frezier, *Detroit do Magellan et malouines,* 1716. Repository: National Library of France, Charts and Maps Department, Digital.
41. J. Bellin, *Carte réduite de la partie la plus meridionale de l'Amerique,*1740. Repository: Institut Cartogràfic i Geològic de Catalunya.
42. J. B. D'Anville, Carte de Mr. D'Anville, 1748. National Historical Archives of Spain, State Section. Maps, Plans and Drawings, sig. 80. Repository: National Library of Spain. Digital.
43. F. Ruiz Puente, *Plano de las Islas Malvinas*, 1769. Repository: National Library of Spain. Digital.
44. J. Bellin, *Carte Reduite des Isles Malouines ou Isles Nouvelles,* 1771. Repository: National Library of France, Charts and Maps Department, Digital.
45. D. Pernetty, *Plan of Acarron Bay situated at the East Point of the Malouine Islands,* 1771, John Carter Brown Map Collection.
46. J. Bellin, *Carte reduite des Isles Malouines ou Isles Nouvelles*, 1771. Repository: National Library of France, Charts and Maps Department, Digital.
47. J. Cook, *A Chart of the Southern extremity of America,* 1775. Repository: David Rumsey Map Collection, David Rumsey Map Center, Stanford Libraries.

48. The results of the survey carried out by James Colnett, published in London in 1798 with the title *A voyage to the South Atlantic and Round Cape Horn into the Pacific Ocean: Fort the Purpose of Extending the Spermaceti Whale Fisheries and Other Objects of Commerce* […] offered the most complete guide for the transit of the English expeditions to the southern region (Flores Guzmán 2010).
49. T. Jefferys, *South America with the adjacent Islands in the Southern Ocean and South Sea,* 1776. Repository: David Rumsey Map Collection, David Rumsey Map Center, Stanford Libraries.
50. J. B. D'Anville, *A map of South America,* 1794. Repository: David Rumsey Map Collection, David Rumsey Map Center, Stanford Libraries.
51. R. Wikinson, *South America,* 1806. Repository: David Rumsey Map Collection, David Rumsey Map Center, Stanford Libraries.
52. J. Cary, South America south, 1807. Repository: David Rumsey Map Collection, David Rumsey Map Center, Stanford Libraries.
53. Harley (2005: 141–184) distinguished between the internal and external power of cartography in order to underline the overlapping of knowledge and power in the production processes themselves. He explained how the compilation, generalization, classification, hierarchical organization and standardization of geographical data are far from being simple neutral technical tasks and that, upon exercising the internal power, cartographers responded to the dictates of external power, even when it seemed feeble.
54. T. Kitchin, *A new map of the world: with all the new discoveries by Capt. Cook and other* […], 1799. Repository: David Rumsey Map Collection, David Rumsey Map Center, Stanford Libraries.
55. Report of Captain Joaquín Olivares y Centeno n/d, attached to Francisco Orozco's official communication to Julián de Arriaga dated 16th October 1759, Royal Palace Library (Madrid), Ms. 2820, fs. 343–347.
56. Governance of the Malvinas Islands, *Plano de las Yslas Malvinas (situadas en la latitud de 51 grados, 28 min. Sur y en 316 grados y 30 minutos según del meridiano de Tenerife, su longitud) arreglado a los últimos reconocimientos hechos en ellas por mar y tierra de orden de su Gobernador Dn. Phelipe Ruiz Puente en el año 1769.* Repository: National Library of Spain. Digital.
57. State account elaborated by His Excellency Mr. Manuel Amat y Junyent, Viceroy of Perú and Chile to his successor His Excellency Mr. Manuel de Guirior, comprehensive from 12th October 1761 until 17th July 1776. National Library (Madrid) Ms. 3111, page 370 (back) – 376. The Viceroy justified an annual cost of over 60.000 pesos, explaining: "Said islands are extremely cold, and barren, but it is essential to support this settlement, in order to prevent foreign nations from making a scheduled stop, and easily get into these southern seas".
58. Gálvez a Vértiz, Aranjuez 8th June 1778, *Apuntes y advertencias para las Instrucciones que se deben formar en Buenos Ayres por el Virrey de aquellas Provincias* […], General Archive of the Indies, Audiencia de Buenos Aires 326.

59. Vértiz to Gálvez, Montevideo 30th April 1781 and Vértiz to Gálvez, Montevideo 24th December 1781, General Archive of the Indies, Audiencia de Buenos Aires 327.
60. J. de la C. Cano y Olmedilla, *Mapa Geográfico de América Meridional*, second edition, 1776. Copy filed with the National Historical Archives, Madrid, et seq 676 and 738, 321 and 328. Repository: David Rumsey Map Collection, David Rumsey Map Center, Stanford Libraries.
61. J. de la C. Cano y Olmedilla, *Mapa Marítimo del Estrecho de Magallanes,* 1769. We use T. Jefferys's edition, 1776. Repository: David Rumsey Map Collection, David Rumsey Map Center, Stanford Libraries. Image in Chapter 4.
62. J. C. Sáa y Faría, *Mapa Geográfico que comprende todos los modernos descubrimientos de la Costa Patagónica y sus Puertos desde el Río de la Plata hasta el Puerto del Río Gallegos, junto al Cabo de las Vírgenes; la porción descubierta del Río Negro y caminos por la campaña desde Buenos Aires. El cual mandó formar por las memorias y planes adquiridos el Exmo. Señor Marqués de Loreto Virrey Gobernador y Capitán General de estas Provincias del Río de la Plata. Año 1786*. Reproduced by Torre Revello 1938. Digital repository: Europeana.
63. Sáa y Faría to Marqués de Loreto, Buenos Aires, 2 September 1786, General Archive of the Indies, Buenos Aires 99.
64. *Carta esférica de las costas de América Meridional desde el paralelo 36 grados de latitud Sur hasta el Cabo de Hornos, levantada por orden del rey, en 1789, 90, 94 y 95 por varios oficiales de su real armada, presentada a su Majestad por mano del Excelentísimo Señor Don Juan de Langara, Secretario de Estado y del Despacho Universal de Marina. Año de 1798.* Madrid Naval Museum Archive.
65. The journals and other expedition materials were published in nine volumes by Madrid Naval Museum (Various authors 1990–1999).
66. The documents consulted involved Jesuit materials, resulting from the Domingo Perler's surveys (1768), reports of James Cook (1771–1775) and Antonio de Córdoba (1785–1788) and several documents provided by colonial officers in Río de la Plata and Chile. In his journal, Malaspina states having received from Sáa y Frías written and cartographic documents that included the results of the last reconnaissance of the Patagonian shores.
67. Although the military operation carried out in 1806 and 1807 in order to take Montevideo and Buenos Aires had no authorization from the government, the Commander H. Pophan was not a naval adventurer. He thoroughly knew the political and economic benefits of the Enterprise, since the idea of undermining the Spanish Empire by conquering its colonies overseas had been present in the British Cabinet discussions, at least since 1804 (Ferns 1984: 30–31).
68. W. Faden, *Chart of the Coasts of South America from Rio de la Plata to Cape Horn, and from Cape Horn to Valparayso, including the Isles of Juan Fernández,* 1802. Repository: John Carter Brown Map Collection

69. M. de Lastarria, *Colonias orientales del Río Paraguay o de la Plata.* Documents for the Argentinian History, Philosophy School, Buenos Aires, 1914, Vol. III. Digital repository: University of Toronto, Digital. Image in Chapter 5
70. For information on these plans: "*Diario de una expedición a Salinas emprendida por orden del Marqués de Loreto virrey de Buenos Aires en 1786 por Don Pablo Zizur*", "*Diario de un reconocimiento de las guardias y fortines que guarnecen la línea de frontera de Buenos Aires para ensancharla por Don Félix de Azara*" (1796); "*Instrucciones*" from Luis de Alaba to Luis de la Cruz and Journal (1806) and "*Examen crítico del diario de Don Luis de la Cruz por una comisión del Consulado de Buenos Aires y defensa del autor*", documents published in De Angelis' collection (1972).
71. As Penhos (2005) notes, there is a common thread between the illuminated medieval manuscripts known as books of wonders, copied by hundreds, and the journey accounts that were printed and became an editorial boom in the 18th century, since both brought news from farther places.

References

Andrews KR (1968) The Aims of Drake's Expedition of 1577-1580. The American Historical Review 73(3):724–741

Andrews KR (1981) Beyond the equinoctial: England and South America in the sixteenth century. The Journal of Imperial and Commonwealth History 10(1):4–24

Andrews KR (1982) On the Way to Peru: Elizabethan Ambitions in America South of Capricorn. Terrae Incognitae 14(1):61–75

AAVV (1990–1999) *La Expedición Malaspina 1789-1794*, Barcelona/Madrid, Ministerio de Defensa/Museo Naval/Lunwerg

Barros JM (1978) El primer testimonio de Tomé Hernández sobre las fundaciones hispánicas del Estrecho de Magallanes. Anales de Instituto de la Patagonia, Punta Arenas 9:65–75

Boungainville, Louis-Antoine (2005) [1° ed. 1771] *Viaje alrededor del mundo a bordo de la fragata real la Boudeuse y la urca Étoile en 1766, 1767, 1768 y 1769*, Buenos Aires, Museo del Fin del Mundo - Eudeba

Bradley PT (1989) *The lure of Peru: maritime intrusion into the South Sea, 1598-1701*. Londres, McMillan

Buisseret, D. (2007) "Spanish Colonial Cartography, 1450–1700", Woodward, D. (ed.), *Cartography in the European Renaissance - The History of Cartography,* vol. 3, Chicago, University of Chicago Press

Cerezo Martínez R (1994) *La cartografía marítima española en los siglos XIV*. XV y XVI, Madrid, Consejo Superior de Investigaciones Científicas

Chaunu P (1984) *Conquista y explotación de los nuevos mundos (siglo XVI).* Labor, Barcelona

Crespo Solana A (2013) Las rivalidades hispano-neerlandesas en el Pacífico y la conquista de Australia: de Cornelis de Houtman a Abel Janszoon Tasman (1595-1651). Anuario de Estudios Americanos 70(2):479–507

De Angelis P (1972) *Colección de obras y documentos relativos a la Historia Antigua y Moderna de las Provincias del Río de la Plata.* T. VIII A, Buenos Aires, Plus Ultra

Destombes M (1955) The Chart of Magellan. Imago Mundi 12:65–88

Ferns HS (1984) *Gran Bretaña y Argentina en el siglo XIX*. Buenos Aires, Solar

Fitz Roy R (2013) *Viajes del Adventure y el Beagle.* Diario, Madrid, CSIC

Flores Guzmán RA (2010) Los balleneros anglonorteamericanos y la apertura comercial del Pacífico sur a fines de la época colonial (1790-1820). Histórica 34(2):63–98

Frost A, Williams G (1997) The Beginnings of Britain's Exploration of the Pacific Ocean in the Eighteenth Century. The Mariner's Mirror 83(4):410–418

García Insausti J (2017) Para la propagación de la fe, para el servicio del rey, y utilidad de sus estados y vasallos. Proyecto anónimo para la instalación de una misión en las Islas Malvinas alrededor de 1766. Antiguos jesuitas en Iberoamérica 5(1):208–226

Goebel J (1982) *The struggle for the Falkland Island: a study in legal and diplomatic history.* Yale University Press, New Haven

González Leiva, J. I. (2004) "La expedición Malaspina y la cartografía sobre Chile", en *Revista de Geografía Norte Grande,* 31: 7–29

Haesbaert, R. (2007) El mito de la desterritorialización: del "fin de los territorios" a la multiterritorialidad, México, Siglo XXI

Haesbaert, R. and Limonad, E. (2007) "O território em tempos de globalização", *Revista Electrônica de Ciências Sociais Aplicadas,* Vol. 1, 2 (4): 39–52

Keuning J (1949) Hessel Gerritsz. Imago Mundi 6:48–66

Kitchin, R. y Perkins, C. (2009) "Thinking about maps", Dodge, M., Kitchin, R. and Perkins, C., *Rethinking Maps,* Londres, Taylor & Francis

Lacoste P (2003) *La imagen del otro en las relaciones de la Argentina y Chile (1534-2000).* Buenos Aires, Fondo de Cultura Económica

Laguarda Trías R (1994) El descubrimiento de las Islas Malvinas en 1520 y su predescubrimiento presunto. Revista de Historia de América 118:51–81

León Solís L (1994) Los araucanos y la amenaza de ultramar, 1750-1807. Revista de Indias, LIV 20:313–354

Levaggi A (1978) El presidio español de las Islas Malvinas. Investigaciones y Ensayos 24:351–389

Lois C (2018) Terrae incognitae. Modos de pensar y mapear geografías desconocidas, Buenos Aires, Eudeba

Lozano, P. (1969) "Diario de un viaje a la costa de la Mar Magallánica en 1745 desde Buenos Aires hasta el Estrecho de Magallanes, formado sobre las observaciones de los padres Cardiel y Quiroga", De Angelis P., *Colección de obras y documentos,* T. III, Buenos Aires, Plus Ultra, pp. 593–633

Luiz MT (2006) *Relaciones interétnicas en Patagonia.* HANIS-Universidad Nacional de la Patagonia San Juan Bosco, La convivencia hispano-indígena a fines del período colonial

Luiz, M.T. and Gullón Abao A. J. (2014) "El Estrecho de Magallanes en una imagen del final del Antiguo Régimen sobreviviente durante el siglo XIX", Cuesta Domingo, M. (director and editor), *Cartografía hispánica (1800–1975),* Madrid, Ministerio de Defensa

Luiz, M. and Schillat, M. (1997) *La frontera austral. Tierra del Fuego 1520–1920,* Cádiz, Servicio de Publicaciones de la Universidad de Cádiz

Maroto Camino M (2005) *Producing the Pacific: Maps and narratives of Spanish exploration (1567-1606).* Nueva York, Rodopi

Martinic M (1997) Rarezas cartográficas. I. Las cuatro versiones del mapa de Chile del padre Alonso de Ovalle. II. El curioso mapa de la región magallánica de Francisco de Seixas y Lovera (1690). Boletín de la Academia Chilena de la Historia 107:385–400

Martinic M (1999) *Cartografía Magallánica 1523-1945.* Universidad de Magallanes, Punta Arenas

Martinic M (2016) Bucaneros en el estrecho de Magallanes durante la segunda mitad del siglo XVII, nuevos antecedentes. Magallania 44(1):5–14

Navarro Floria P (1994) *Ciencia y política en la región norpatagónica: el ciclo fundador (1779-1806).* Universidad de la Frontera, Temuco

Nocetti, O.R. and Mir, L. B (1997) *La disputa por la tierra. Tucumán, Río de la Plata y Chile, 1531–1822,* Buenos Aires, Sudamericana

Noejovich HO, Salles EC (2011) La defensa del Virreinato del Perú: aspectos políticos y económicos (1560-1714). Fronteras de la Historia 16(2):327–364

Padrón R (2009) A Sea of Denial: The Early Modern Spanish Invention of the Pacific Rim. Hispanic Review 77(1):1–27

Padrón R (2010) "Las Indias olvidadas: Filipinas y América en la cartografía imperial española", *3º Simposio Iberoamericano de História da Cartografia*, Universidade de São Paulo

Penhos, M. (2005) *Ver, conocer, dominar: imágenes de Sudamérica a fines del siglo XVIII*, Buenos Aires, Siglo XXI

Pernetty D (2012) [1770] *Historia de un viaje a las islas Malvinas*. Buenos Aires, Museo del Fin del Mundo - Eudeba

Pigafetta A (1963) *Primer viaje en torno al globo*. Espasa-Calpe, Madrid

Porro Gutiérrez, J. M. (2003) "Una antinomia protorrenacentista: secreto de estado y divulgación en los descubrimientos luso-castellanos", *Anuario de Estudios Americanos*, t. LX, 1: 13–40

Raffestin C (2011) *Por una geografía del poder*. México, El Colegio de Michoacán

Ratto HR (1930) *Actividades marítimas en la Patagonia durante los siglos XVII y XVIII*. Buenos Aires, Kraft

Rose JH (1929) Captain Cook and the Founding of British Power in the Pacific. Geographical Journal 73(2):102–122

Sáa y Faría, J. C. de (1972) "Respuesta del Brigadier don José Custodio de Sáa y Faría a S.E. sobre el reconocimiento y diario de Villarino" [1783], P. de Angelis, *Colección de obras y documentos*, T. VIII B, Buenos Aires, Plus Ultra, pp. 1147–1150

Sánchez, A. (2017) "D'une carte inexistante à un empire colonial. La cosmographie de la Renaissance sous la Monarchie espagnole à partir du Padrón Real", *L'Atelier du Centre de recherches historiques. Revue électronique du CRH* (17)

Sankey, M. (2003) "The French in Terra Australia", *Arts: The Journal of the Sydney University*, 25: 26–63

Schmidt B (1999) Exotic allies: The Dutch-Chilean encounter and the (failed) conquest of America. Renaissance Quarterly 52(2):440–473

Schouten, W. (2010) *Diario o descripción del admirable viaje de Willem Schouten, holandés. De cómo descubrió un nuevo pasaje, hasta ahora desconocido, hacia el sur del estrecho de Magallanes para llegar al Mar del Sur*, Buenos Aires, Eudeba/ Colección Reservada del Museo del Fin del Mundo

Schurz WL (1922) The Spanish Lake. The Hispanic American Historical Review 5(2):181–194

Spate, O. H. K. (1979) *The Spanish Lake. The Pacific since Magellan*, vol. 1, Minneapolis, University of Minnesota Press

Spate (1983) *Monopolist and freebooters. The Pacific since Magellan*, vol. 2, Minneapolis, University of Minnesota Press

Torres Santo DM (2004) "Un bestseller del siglo XVIII: el viaje de George Anson alrededor del mundo", *Biblio 3W*. Revista Bibliográfica de Geografía y Ciencias Sociales 9(531):1–27

Urbina CMX (2016) La sospecha de ingleses en el extremo sur de Chile, 1669–1683: actitudes imperiales y locales como consecuencia de la expedición de John Narborough. Magallania (Chile) 44(1):15–40

Vega Palma, A. (2006) "Representación cartográfica de la Gobernación de Chile en el siglo XVI: la cordillera y la construcción de la identidad territorial", Lois, C. and others, *Imágenes y lenguajes cartográficos en las representaciones del espacio y del tiempo*. I Simposio Iberoamericano de Historia de la Cartografía, Instituto de Geografía - Facultad de Filosofía y Letras - Universidad de Buenos Aires

Wallis, H., (1965) "English Enterprise in the region of the Strait of Magellan", Parker, J. (Ed.), *Merchants & Scholars. Essays in the history of exploration and trade*, Minnesota, University of Minnesota

Wallis H (1978) Publication of Cook's Journal: some New Sources and Assessments", *Pacific Studies*, 1 (2): 163–294

Wallis H (1984) "The Cartography of Drake's voyage", Thrower, N. W. (ed.), *Sir Francis Drake and the famous voyage, 1577–1589*, Los Angeles, University of California

Chapter 4
Cartographic Representations of Indigenous Territoriality

Abstract The debate on Patagonia as a useful (less) place re-created views of an inhospitable land inhabited by savages. Instead, cartography and Jesuit accounts depicted a land inhabited by people who built their territory in accordance with their own interests. We delve into the production conditions of maps, strategies for collecting information, indigenous knowledge interpretation and the harmonization of different spatial conceptions. We examine the imperial use of Jesuit cartography when exploring the interior of Patagonia, border policies design and the construction of a cartographic discourse that contributed to the intellectual and symbolic appropriation of space. We consider other sources that evince native informants' participation in mapmaking and provide elements to access to the territoriality of the peoples of the southernmost end.

Keywords Indigenous territoriality · Jesuit and imperial cartography · 17th, 18th and 19th century

4.1 Introduction

Historical cartography studies show that the exclusive attention paid to the necessary data for navigation and the provision of fleets in interoceanic transit can explain the progress in geographical knowledge of the southern region (Martinic 1999; Porro Gutiérrez 1999). Those found in fantasy iconography, re-creating myths and European imaginary beliefs—City of the Caesars and the "Patagonian giants"—(Luiz and Schillat 1997) or else insisting on cartographic blanks, were the only references to the interior of Patagonia and Tierra del Fuego. This, as Harley (2005) remarked, promoted the idea of a socially deserted place.

Jesuit cartography constitutes an exception, by incorporating information provided by natives, as it enabled significant broadening of knowledge of the continental interior still unexplored by Europeans.

Cartographic work of missionaries on the southern border of Chile and Río de la Plata fulfilled a double purpose, political-institutional and practical. On the one hand, the Order's interest to account for the inclusion of territories is considered by the

L. I. de Lasa and M. T. Luiz, *The Southernmost End of South America Through Cartography*, The Latin American Studies Book Series,
https://doi.org/10.1007/978-3-030-65879-3_4

European imaginary as "remote, unreached and inhospitable edges of the empire" in their labor; on the other hand, the need to solve spatial issues with a view to settle missions in an unknown territory is controlled by natives. Maps drawn in the 18th century provide accurate references on aspects connected to the instrumental use of cartography, namely location, localization and distances, habitability conditions, resources, territory occupation, strategic sites, routes and borders. Likewise, they convey certain information that, as Pickles (1992 in Quintero 2000: 196) pointed out, "has been selected in order to support an argument". In this case, they were used to demonstrate the possibility of incorporating an inhabited and habitable territory as an evangelization field.

Taking into account the interests at stake in the production of Jesuit cartography and its use by members of the Society of Jesus and later by colonial agents, it is worth highlighting its propositional intent. Maps not only describe a territory by transmitting situational information[1] but are also produced with a certain intention. They "propose a space to act un-covering possibilities and anticipating effects" (Kitchin and Perkins 2009).

The maps of José Cardiel and Thomas Falkner, followers of the cartographic work initiated during the 17th century by Alonso de Ovalle and the Jesuits of the Chile Province, represent a "local situation" that merges native lore into the European referential systems. By doing so, they attempt to harmonize different spatial conceptions—experience-based and abstract space—unveiling other ways of viewing and experiencing the territory.

From the most interesting feature—the prominence of native sources of information in map production—we observe the specificity of the border space as an inter- and intra-ethnic relationship network. It was within such a network where the fragmentary, irregular and intermittent nature of exchanges, the complexity of contact situations and the uncertainty of communication demanded the production of several procedures to obtain, validate and select information. This process not only implies the efforts to construe and decipher what is transmitted and what remains concealed but also, as Belyea (1992) states, the difficult task of fitting the given data into the geographical scheme and European cartographic conventions or, more precisely, the creation of a series of equivalent conventions that describe spatial characteristics.

Bearing in mind the influence of Jesuit production on western cartography of the 17th and 18th centuries (Furlong Cardiff 1936; Udías 2015), it is worthwhile highlighting the central role of missionaries. They were intermediaries between native and European geographical knowledge, protagonists—together with local informants—of different ways of perceiving, representing and valuing the territory. In a way, they are two different cartographies (Fig. 4.1).[2]

Missionaries' intellectual and technical training,[3] their observational and filing skills and their aptitude for communicating with native inhabitants enabled them to collect and give meaning to their informants' information, contrasting testimonies and making use of their own observations (Di Liscia and Prina 2002).

The range of information collected in the missions and in their exploratory trips guided by experts and its systematization in letters, accounts and maps enable us to consider it as the first knowledge corpus on southern territories in America.

We examine the general maps of Patagonia produced by Alonso de Ovalle (1646), José Cardiel (1746–1747, 1749 and 1751) and Thomas Falkner (1772) within the framework of the Order's mapping and expansion plan. We consider their imperial utilization in the exploration of the interior, the occupation and territorial control strategies designs and as a source of official cartographies. We focus on the production of Basilio Villarino (1782–1783) and José Custodio Sáa y Faría (1786), due to the central role assigned to the indigenous spatial information, obtained directly from local guides or informants and/or selected from the Jesuit corpus and from other colonial agents' materials. Finally, we are interested in introducing several sources that provide access to knowledge on peoples' territoriality in the southern end and that evince the participation of native informants (Fig. 4.2).[4]

4.2 Patagonia as a Social Space: Ovalle's Story Map

Alonso de Ovalle's *Tabula Geographica Regni Chile*, chart appearing in "*Histórica Relación del Reino de Chile*", published in Rome in 1646, includes southern territories from 25° south to Tierra del Fuego.[5] Based on Jean de Laet's map (1625), with information collected by the author and other Jesuits who worked in the missions in the south of Chile, it represents four territories as integrating the Captaincy of Chile. Those are Chile itself and Cuyo, known by the Spaniards, and the Magellan region and the Strait of Magellan—including Tierra del Fuego archipelago—just known by the natives.

Father Furlong Cardiff's views (1936) regarding the vagueness of ideas and the ignorance in respect to the interior of the continent conditioned the reading of Ovalle's map.[6] Nevertheless, the cartographic image shows significant progress in knowledge

Fig. 4.1 Ovalle Tabula Geographica Regni Chile (1646)

Fig. 4.2 De L'Isle Carte du Paraguay, du Chili, du Detroit de Magallan (1703)

organization when it comes to Patagonian indigenous territories, mainly in the depiction of the hydrographic network. Almost all the largest lakes and the sources of most important rivers are found in the west hydrographic gradient, whereas the depiction of the east gradient necessarily required information provided by native informants, since the area was still unexplored. Only the river mouths could be attributed to the European knowledge of the Atlantic coast. Thus, the sources of certain rivers are depicted—Negro, Deseado and Santa Cruz—while others—Colorado and Limay—are simply omitted. Taking into account that the information provided by Flores de León's expedition (1620) and Father Rosales' references (1653) to the "famous Nahuel-Huapi lagoon" (Furlong 1992: 40) were available, the lack of details probably accounts for the cautious presentation of the uncertain data available. Moreover, it could result from a strategic attitude of the Amerindians, concealing or twisting information on key areas in order to control their territory (de Lasa and Luiz 2011) (Fig. 4.3).[7]

As Hanisch has noted (1976: 228), illustrations play a prominent role on Ovalle's map, since they bolster or clarify the content of *Histórica Relación*. Without denying the cartographic representation and written work complement each other, we observe that illustrations supply the map with a story that can be construed independently from the text itself. Drawings represent regional context elements: autochthonous fauna (guanaco, ostrich), horses in the *pampas* (grassland) and native inhabitants wearing their own clothing. Spanish settlements were correctly identified on the western slope of the Andes and, on the Atlantic coastline, next to the Magellan Strait, the ephemeral city of Rey San Felipe was indicated with a figure of a ploughman. The fabulous beings were located in the interior of Tierra del Fuego and in the oceans, which denotes the complete lack of information as regards the lands to the south of the Magellan Strait and the perils of the sea in these southern latitudes.

The pictures help us identify the coexistence of information provided by Spanish and Amerindian sources and that produced by the European imaginary, such as the representation of an encounter between seamen and giants on the eastern coast of Tierra del Fuego and that of a tailed man and a woman covered in mud in the west of the island.[8]

With these symbolic elements in mind, it is noticeable that the Chilean Jesuit's chart contests the image shaped by the European incipient state of knowledge, strongly influenced by the perception of distance and harsh weather. The 17th century cartographic discourse re-created a deserted and desolated region by means of "blanks", whereas Ovalle's map led to a different interpretation and other estimations. Serving the Order's purpose, the account and map *produce a space for action*, by representing a territory with history, inhabited and habitable, and therefore, suitable for becoming part of the missionary project. Furthermore, the work portrays the existence of resources of interest, such as cinnamon and pepper, and the strategic advantages of the passage to the Pacific Ocean, reasons that could be the grounds of the imperial decision of founding settlements and missions in the Magellan region (Burdick 2014).

The legend or "note" included in the re-edition of Ovalle's map that was published together with Father Diego de Rosales' *Historia General del Reino de Chile, Flandes Indiano* confirms the Order's intention to promote the region as a potential evangelization field. At the beginning of the book, a text reinforces the iconic register on Ovalle's map. It refers to natural riches and weather conditions: "*You will see here the most pleasant and fertile land in America according to Laet. The most copious domain in metals and of the healthiest temperament in the view of Antonio de Herrera. The most similar region to Spain as Bry and Fr. Gregorio de León affirm*" (Furlong Cardiff 1936: 32) (Fig. 4.4).[9]

Ovalle started a compilation and systematization cycle of information provided by natives, continued by Nicolás Mascardi, Bernardo Havestadt, Diego Rosales, José Quiroga, José Cardiel and Thomas Falkner.

During the second half of the 17th century and the first half of the 18th century, Jesuits made several exploratory trips across indigenous territories on the eastern slope of the Andes, though they never produced general maps of the area. Bernardo Havestadt's map, printed in 1752 and appearing in the book *Chilidungun*, is a small-scale chart of the central region of Chile, south of Mendoza and north of Neuquén. These areas had been visited by the end of 1751 and the beginning of 1752, in the context of the "Circular Missions" or itinerant, method adopted due to the difficulties of settling regular missions among trans-Andean groups (Villalobos 1989; Barcelos 2007). The interpretation of the map, which does not depict key elements of the territory such as indigenous routes or the Andes passes, requires reading the journey log. Havestadt's account, written in order to provide details for future experiences, offers a thorough description of the space and visited places (Brañes 2006).

Another partial map of the area is the one produced by Father José Quiroga in 1745, entitled *Costa de los Patagones*.[10] It contains the results of the Atlantic coast reconnaissance carried out during Joaquín de Olivares' expedition, organized by missioner Mathias Strobel, and in which José Cardiel took part.[11] The chart does not provide new information on the interior, just a few indigenous toponyms in the sites of the missions to the south of the Salado River.

The Jesuit corpus constituted one of the major sources of French and Spanish cartography (Furlong Cardiff 1936). Nicolás Sanson d'Abbeville used Ovalle's map to elaborate different maps of the southern region of America between 1650 and 1679 (Martinic 1999). By adapting the selected data to modern features of the French School, the author attained a simpler cartographic image, which facilitates its reading, modality still used by Jesuit cartographers in the 18th century.

The *Carte du Paraguay, du Chili, du Detroit de Magellan* (1703) by Guillaume De L'Isle, a map considered a cartographic model for half a century, also uses Ovalle's work for southern territories, though it omits depicting river courses in southern Patagonia, possibly due to lack of data confirmation by new European reconnaissance. De L'Iste indicates the territorial nature of *Terre Magallanique* by means of illustrations that appear next to the cartouche on one side of the map. This being a couple of *tehuelches* covering their trunks with blankets (European?) and a scene of naked canoeists with slim bodies and postures like the classical Greek-Roman iconography, appearing in the inferior sector, at the insular end of the continent. The ornamental figure enhances the stereotype of Patagonian native inhabitants that persisted into the beginning of the 19th century. Such stereotypes portray the native inhabitants of northern Patagonia as civilizable "barbarians", capable of adopting not only western clothing but also western habits, in contrast with the "savage" nomads, whose otherness, difficult to accept and understand, fades when depicting the characters with European features.[12]

When drawing their maps, D'Anville, Bellin and Cano y Olmedilla took into account the improvements in the localization of rivers and ports that Quiroga had achieved in his own map. D'Anville's 1748 *Amerique Meridionale*[13] includes information as regards the Patagonian interior taken from Jesuit sources, utilizing indigenous toponyms, locating the Nahuel Huapi mission, founded by Father Mascardi in 1669 and destroyed in 1717, and the Concepción de los Pampas mission, established in 1740 in the south of Buenos Aires province.

D'Anville, *Amerique Meridionale* (1748)

http://bdh.bne.es/bnesearch/CompleteSearch.do?field=todos&text=anville&showYearItems=&exact=on&textH=&advanced=false&completeText=&pageSize=1&pageSizeAbrv=30&pageNumber=57

4.3 Integrating Natives' Spatial Knowledge in Cardiel and Falkner's Maps

The southern region's "double frontier" nature—contact area with independent native inhabitants and space for convergent interests of different European states—determined the scope of Patagonia's peopling and the identity of its settlements. Foreign and indigenous threats of invasion compelled broadening the knowledge of the continent, in view of the design of a new border policy aimed at preventing an occupation by rival states through alliances with native inhabitants. Moreover, they needed to control and restrain movements of groups that were monitoring the routes and frontier areas, creating a space for consensus to favor collaboration.

The resistance of the Araucanía, Pampa and Patagonia nations delimited the missionary policy developed by the Society of Jesus from Valdivia and Chiloé to

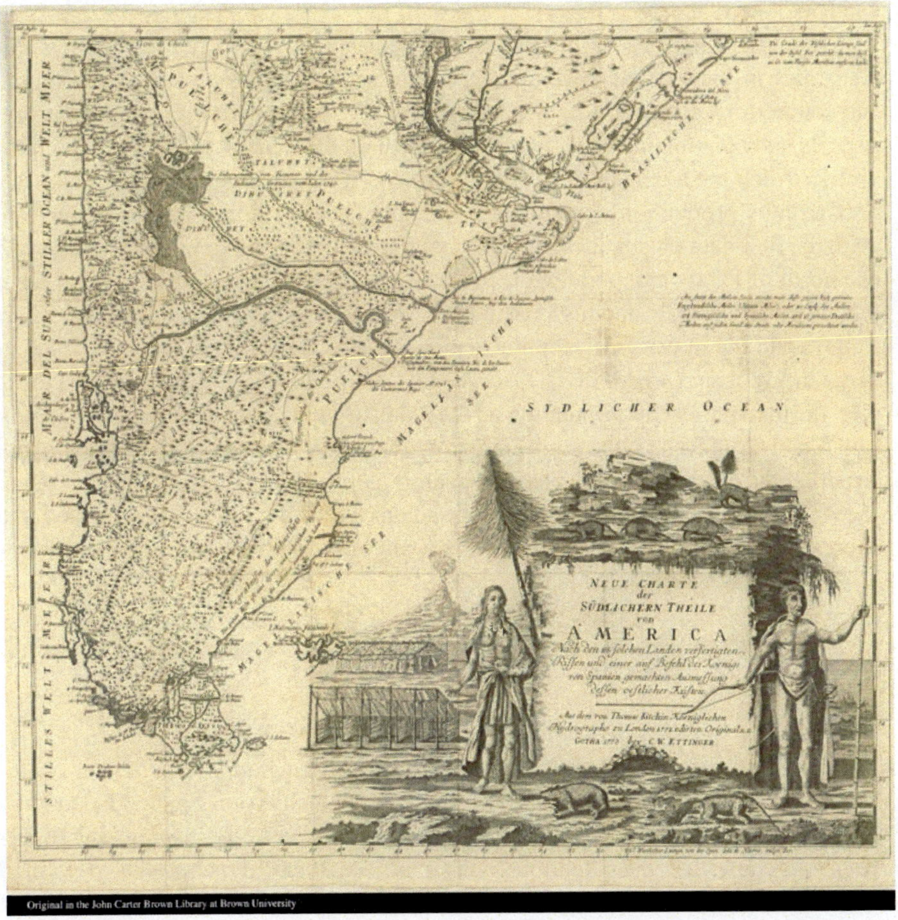

Fig. 4.3 Falkner *A new map of Southern Parts of America* (1772)

the trans-Andean region in Neuquén, as well as in the Río de la Plata to the south of the Salado River.[14] Likewise, the construction of an indigenous space structured on a vast network of exchanges that connected the Pampa region to the territories in the center-south of Chile, revitalized during the 18th century by cattle trading and strong links with the colonial market[15] framed its scope.

From *Nuestra Señora del Pilar del Volcán* reduction, devoted to the groups living in the *serrana* region to the southwest of Buenos Aires province up to the Negro River, José Cardiel and Thomas Falkner carried on with the cartographic labor initiated by the Jesuits of Chile Province, broadening the knowledge of the Patagonia interior by incorporating native spatial information.

We delve into three of Cardiel's maps, which represent southern territories in Paraguay Province—*Carta de la extremidad austral de América* (1747),[16] *Tierra de Magallanes* (1749)[17] and *Mapa de Magallanes* (1751)[18]—and the map published together with the account of Falkner's voyage. The latter produced in London in 1772 by engraver Thomas Kitchin with information compiled by the missionary in the Río de la Plata.[19]

These maps may be deemed as different interpretative versions of the territorial dynamics in view of the development of an evangelization project that comprised the southern region up to the Magellan Strait.[20] To this end, the focus is on social aspects and information selection leads to explain them, particularly the depiction of rivers, which are crucial to understand the economic functioning of the region.

Cardiel's production constitutes a valuable source since it provides the Jesuits with the first data on spatial distribution of population, seasonality and resources. In the north of Patagonia, population concentrates on the Andes slopes, Malarhue and Maule gulley, middle and high basin of the Neuquén River, Limay and Collón Curá valleys and the lower valleys of the Negro and Colorado rivers. The most inhabited areas in Araucania are Valdivia, Osorno and Llanquihue. In the southernmost territories, the map represents population on the plateaus of the middle basin of the Deseado and Santa Cruz Rivers, the Andean region, the grassy steppes close to the Magellan Strait and Tierra del Fuego. The Pampean region is not permanently inhabited and the Atlantic coast, "deserted", "barren" and uninhabited all along, is neglected in the representation, since it is ruled out as an evangelization field.

Cardiel—Palomares, *Mapa de Magallanes* (1751)
http://bdh-rd.bne.es/viewer.vm?id=0000143616

Numerical references and notes on *Mapa de Magallanes* (1751), as well as symbols showing different types of settlements—cities of Spaniards, towns of Indians, *tolderías* (camps of Indian huts) or *habitaciones de infieles* (unbelievers' rooms)—show that the attention focuses on the social dynamics in view of the construction of a missionary field. Considering the location of indigenous settlements, it is clear that the representation matches with the characterization provided by ethno-historical studies of "a bit dispersed pattern albeit considerably uniform and with variable density in relation to the areas and resources availability" (Palermo 1988: 87). Upon identifying "on-horse nations"—Araucanía, northern Patagonia and southern Buenos Aires mountains—and those of "Indians on foot"—southern Patagonia, it acknowledges that using horses constitutes a key element to understand the functioning of the territory. The *Carta de la extremidad austral de América* (1747), which presumably inspired the previous map, differentiates the areas through coloration. It also portrays two vast areas of population distribution: the Pampa as an area for cattle provisioning and the areas all along the coast in white, showing them as "deserts", less suitable for settlements.[21]

Thus, cartographical images provide elements to approach the different ways in which maps build and communicate socioeconomic dynamics knowledge, by merging two spatial representations: the Indigenous one, which focuses on differentiating areas and lines of movement, and the European one, which situates the different elements as fixed spots. Such areas constitute the several nations, not separated by accurate boundaries but by fluctuating transition areas; lines of movement account for these peoples' itineraries, achieved "through complex memorization of distinct details sequence" (Lynch 1984, 147), conspicuous features of the landscape—mountains, lakes, rivers, hunting spaces—and places identified as orientation references when moving across the land. Cacique Bravo's route is a typical itinerary of different groups and one of the main trading routes.

We note that they selected reduction sites on the grounds of the optimal conditions of the area for its population and its importance within the indigenous economic space, since it was a key area for trading, cattle and horse provision and salt harvesting. This becomes apparent in the missionary's work with the foundation of *Nuestra Señora de la Concepción de los Pampas* (1740–1753), *Nuestra Señora del Pilar del Volcán* (1746–1751) and *Nuestra Señora de los Desamparados* (1750–1751).

As far as hydrography is concerned, the representation of the courses of the Colorado River (*de las Barrancas River*) and of the Negro River (*del Sauce River*) is undoubtedly the result of interpreting and validating information provided by native informants. Since the orientation of both rivers is more accurate than on Ovalle's map, their sources are incomplete. Using Quiroga's map enabled a better localization of the mouths of the vast Patagonian Rivers, improving previous maps, also used by Cardiel. It is worth noticing the absence of the courses of the Santa Cruz and Deseado Rivers, drawn on Ovalle's map from their sources—based on indigenous information—up to their mouths, located according to measurements of European seamen. This omission could be explained by the lack of enough information in order to confirm the data or else by the data selection upon interpretation, which is as a result of considering some data relevant for the understanding of the economic space functioning. Another omission quite difficult to understand, considering the knowledge they had on the routes and strategic areas within the indigenous territory, is that of the Huechulafquen Lake—the hub of cattle trading, which had been well represented on Falkner's map.

Both Cardiel and Falkner emphasize the importance of native people's collaboration, who provided references on spatial and social relationships, topography, distribution, characteristics and population dynamics, and also helped as guides in voyages and located places through toponymy.

It is worth mentioning that both solve the longitude issue by privileging the information provided by the Indians. Cardiel acknowledges having used Amadeo Frezier's map (1716) since it was the one better adapted, "according to what I have found out about the relationships between many Indians of these lands, when I was in the Volcán Oriental Hills" (Outes 1940: 13). In his account, Falkner gave grounds for the modification of the map he used as a basis to represent the continent. He argued, "I was forced to stretch the country from east to west compared to the size it had on D'Anville's map, since it is impossible to merge what appears on such map with the information provided by the Indians or with what I could observe by myself as regards distances between places" (Falkner 1974: 55).

As a rule, Jesuits explicitly stated the validation process used to confer veracity and reliability to the information. Regular communication, facilitated by the fact that missionaries knew native languages and that a growing number of indigenous people knew Spanish,[22] enabled the exchange of services and information required to develop the cartographic work. Challenging different versions became a crucial stage of the process, focusing on the variation margin resulting from interpreters' mediation and the possibility of cheating or concealment of strategic spatial information and, above all, from the difficulties in understanding a system of references and meaning built from a spatial conception that differed from the European one. [23]

A estos [indios de varias naciones] he examinado en diversos tiempos, y sin que uno sepa de otro, para dar mejor con la verdad. Pónese aquí lo más verosímil, y en que todos concuerdan. (Notes taken from *Mapa de Magallanes*, 1751.)

> For the description of the interior, I've normally based myself on my own observations, as I've travelled across most of it, being able to establish places locations, distances and rivers, mountains and jungles. Whenever it was impossible for me to penetrate, I've used the native inhabitants accounts as a source of information, and also what Spanish captives' narrated, since they had lived for years among them and were later rescued (Falkner 1974: 53).

Given the fact that the sense of spatiality differs from one culture to the other one, we wonder about the "process of geographical translation"[24] and about the conditions that helped to overcome conceptual and representational hindrances present in communication.

From the content of the accounts, we can draw the conclusion that collaborative mapping while traveling across the land under the guidance of native inhabitants would have facilitated the complex task of harmonizing the western concept of abstract space with the indigenous concept of space as an inhabited territory. The former is based on geometrical positions and latitude and longitude measurements, while the latter involves the various natural, supernatural, social and political relationships.

Highly mobile societies reached a significant abstraction degree as regards orientation and spatial localization (Lynch 1984), developing, as Ingold remarked (2005), a "sophisticated perceptive awareness of the environment surrounding them and the possibilities of action it offers". These skills allowed for a specialized knowledge, typically reserved to certain members of the group, particularly chiefs.

Hospitality and gift practices, which outlined border relationships in Patagonia (Luiz 2006), allowed for qualified informants' collaboration, whose willingness to cooperate was paramount in the interchange process or, more precisely, in the negotiation of spatial information. Gratification through presents and festivities to honor chiefs and compensation for varied services by means of European goods, which natives incorporated into their preferences, contributed to keeping a system of settlements, concessions and compromises regulating their coexistence and their political and economic transactions.

Missionaries, as well as other colonial field agents traveling across the interior of South America, and unlike most foreign travelers or senior officials, had a keen awareness of the different types of geographical knowledge. Thus, they selected guides and informants based on their experience (Roller 2012). These local experts not only had extensive knowledge of natural conditions, routes and resources but also of socialization protocols among natives that, together with local relationships, assured them the necessary collaboration to travel across territories that were politically fragmented (Jiménez and Aliotto 2016).

In the account introduction, Falkner (1974: 56) revealed the identity of his key informant, by highlighting the service of "great Cacique Cangapol, who lived in Huichin, on the margins of the Negro River", "called by Spaniards the *Cacique Bravo*". Cangapol appears on the frontispiece of the map, as well as his wife Huenec, while in the account the author referred to their intense exchanges: "we were very close and I travelled with him". One of the outcomes of this cooperation during the voyages was the possibility for both, Falkner and Cardiel, to identify the main route connecting the Negro River to the Interserrana area in Buenos Aires, the region in between the Northern and Southern hilly systems that bound it, outlined on their maps as the chief's itinerary.

In the description of the Negro River, Falkner mentions a Cacique cooperating again, providing another undoubtedly valuable piece of information for cartographers.

> A Cacique of the Tehuelches or indigenous people of the south painted 16 of these rivers on a table, naming them all. However, as I did not have anything to write with, I was unable to write them down and I have already forgotten them. He added he did not know any part of the river [Limay], not even before the meeting of these minor courses, as not being very wide and deep. He ignored which could be its source but he was sure it was not in the north (Falkner 1974: 107).

The accuracy of the outline of the upper basin of the river course on Falkner's map when compared with Cardiel's map—on which the orientation is already quite correct—contrasts with the errors and omissions in the representation of other Patagonian hydrographic systems. This could be due to the possibility the author had access to that qualified knowledge and even to observe that ephemeral map drawn on a wooden table by a local chief, who seemed to be "more than seventy years of age" and "had spent all his life by [that] riverside".[25] Apart from depicting the source of the Negro River with two Andean lakes and two Andean passes to Valdivia, the Tehuelche Cacique would have possibly provided information on the river basin at its full length, providing a hydrographical description as well as describing the territorial organization based on cattle trading routes. By doing so, they intended to get the Spaniards alliance to face enemy groups, such as the Araucanians.

Caciques as "intra- and inter-ethnic information processors" (Becchis 1997) selected information while answering questions on issues that were of interest to missionaries: location of bigger settlements, routes and distances, characteristics of the population—whether they were peaceful or warriors—and available resources (watering places, grasslands, wood, fruits, cattle, amongst other things).

Since informants could only offer few references on space that had been inhabited, appropriated and organized in their representation, the territories to the south of the Negro River and some areas of the Pampean region remained virtually unknown. Jesuit cartography evidenced this restraint, in the absence of elements or their correct location or distribution; for example, in the omission of river courses and the excessive number of endorheic basins in southern Patagonia.

Thomas Falkner's map presents an ethnic and demographic image which results coetaneous to Cardiel's, since they collected the information during their stay in the missions in the south of Buenos Aires. Falkner managed to draw one of the most complete maps of Patagonia, based on D'Anville, Quiroga and Cardiel's maps and assessing the information provided by native inhabitants. From a descriptive geographical approach, his work—account and map—supplies plenty of information on the geoenvironmental conditions in the northern Pampa and Patagonia, as well as on its inhabitants' traits. One of the outstanding aspects of Falkner's work is the description and localization of resources of economic interest: fertile lands, surface water, vegetation, salt flats, cattle and horses. It also identifies suitable ports for maritime connections, internal communication routes—rivers and paths—and convenient sites for colonization, highlighting the advantages of peopling the mouth of the Negro River (Bay of Sin Fondo). As regards physical aspects, while the relief—mountains and plateaus—are described and portrayed quite accurately, hydrographic information results incomplete except, as we have mentioned above, in the description and outline of the Negro River. Deseado, Santa Cruz and Gallegos rivers are omitted, contrary to what happens with the middle valley of the Chubut River. In the account, Falkner refers—though hesitantly—to a possible connection between Lake Huechulafquen and the Pacific Ocean.[26]

The absence of southern Patagonia rivers might result from the fact that De L'Isle and then D'Anville do not include them on their maps, and also because Ovalle's information could not be confirmed. Furthermore, it could be due to the insufficiency of available data on the outline of these courses because their local informants did not know the southernmost territories. The lack of information on the Atlantic coastline on other maps, even on D'Anville's, suggests that the interest focused mainly on the interior as a space of missionary work.

As regards the functioning of the indigenous space, the reading of the text and map let us identify territorial boundaries among the several nations, settlements, routes and key areas of the economy. Although we notice that Falkner could not think of the importance of the trading activity as space structuring, the text provides relevant data, such as the description of the way to *Casuhati*, not depicted on the chart (Falkner 1974: 101, 105, 106).

The map enables us to differentiate indigenous and imperial territorialities through toponymy. The former describes the less known interior whereas the latter refers to a space of symbolic control restricted to the maritime coastlines. As regards ethnonyms, it repeats some already mentioned by Cardiel[27] and places them on his map with few variations, reinforcing the coetaneity of the ethnic image. Taking into account the information was elicited from groups inhabiting the mountainous region to the southwest of Buenos Aires up to the Negro River, the lower number of toponyms to the south could result from the fact that their informants did not know the southernmost territories. Despite Falkner mentions that he had contact with a Cacique of the *Yacana-cunnees*, nation of Tierra del Fuego, the information he handles on the archipelago and its inhabitants is quite scarce and reveals the lack of knowledge as regards the interior of Tierra del Fuego when outlining the several channels crossing the Isla Grande. This also applies to the localization of volcanoes, an element that—similarly to what happens with the San Sebastián Channel—is reproduced with very few changes through the 18th century cartography, "filling" the unexplored space.

4.4 Imperial Use of Jesuit Cartography

The analysis of the production conditions of Falkner's map requires considering a series of issues that are difficult to describe. Although, as we already mentioned, the missionary collected and probably systematized the information in the Río de la Plata, but Thomas Kitchin drew the map in England in 1772, with the purpose of publishing it together with the account. Not having the original manuscript of the map nor the text (Livon-Grosman 2003: 62–63), it is impossible to determine the extent of the cartographer and the publisher William Combe's intervention. Likewise, it is not possible to determine either the map referred by Falkner as "my map" (Falkner 1974: 11, 53, 114) is a manuscript of his own or the printed map.[28] Anyway, considering the commercial interest of such a publication as specified in the legend of the map,[29] Kitchin must have prepared the map within the aesthetic and scientific criteria required by consumers. In his work on the trading of maps in France and Great Britain during the 18th century, Sponberg Pedly (2005) explained that maps should be produced from the compilation of topographical surveys, previously published maps, manuscript maps and verbal descriptions of travelers and explorers. Moreover, the title should assure they were based on "the most recent astronomic observations". Another requirement was that one had to be sure of buying a "true" map, the reason why "the cartographer and his reputation were paramount to the trading of maps".

Falkner-Kitchin's map, as Livon-Grosman (2003) accurately referred to it, complies with some of the requirements that guaranteed high quality. On the one hand, its title, "New map of the Southern Parts of America based on manuscript charts performed in the country and a study on the eastern coast commissioned by the King of Spain", and on the other hand, its author, "Thomas Kitchin, Hydrographer to His Majesty", was a well-known engraver in London.

As regards its sources, one of the manuscript charts could be Falkner's himself, considering that the map and the account are complementary; another one could be Cardiel's, having in mind their correspondence in symbols and toponyms. The study on the eastern coast refers to the maritime survey conducted by Jesuits Quiroga, Lozano and Cardiel between 1745 and 1746.[30] The third source, explicitly stated in the text though not on the map, is 1748 D'Anville's chart, a map that amounts to a cartographic model for half a century and, consequently, guaranteed its "veracity".

The map consolidates two conceptions of cartographic representations. It follows D'Anville's aesthetics, his distribution of legends, the frontispiece and the outline of the continent—except for central Patagonia, as stated in the narrative—and, concurrently, it follows Jesuit maps by using symbols similar to Cardiel's, though skipping any references.

Another relevant aspect, considered in several studies on Falkner's work, is its influence within the context of Hispanic-British rivalry. Far from being neutral information, the text informs of the economic and strategic benefits "any power" could achieve from occupying the region, and mentions the vulnerability of southern frontiers. Although the sources of the text were the reports produced for the Order, the lack of religious content and references to the missionary work and the New World controversy, which were characteristic of the expelled Jesuits production (Justo 2015; Martins 2017), evinces a new intention in the use of the material. Assessing southern territories in terms of "utility" in view of the settlement of colonies—and not as an evangelization field—the text calls attention to the convenience of peopling the Bay of Sin Fondo (Falkner 1974: 112–113) and the southeastern coast of Isla Grande of Tierra del Fuego. It is "more effective than the Malvinas Islands in order to control the Southern Sea path" (Falkner 1974: 119–120).

Falkner's map was published in England by in-charge Robert Berkeley, political advisor of the Queen and possible author of the preface, who commissioned the edition of the text to Combe and the production of the map to Kitchin. His aim was to promote the inclusion of Patagonia in the British imperial plans. Spreading this map responded to the interest in promoting the region as a susceptible space to be incorporated to new expansion projects, either private or state-sponsored, on the grounds of scientific, commercial, political and strategic goals. As Fernández Bravo (2004) remarked, "knowledge supplies imperial files and awakes interest in the territory it describes".

The establishment of the missions, though not effective as a strategy of assimilation and integration of territories, made it possible to collect strategic information to assess alternatives of territorial control and to keep peaceful coexistence over the frontiers. Peopling Patagonia was mainly a metropolitan imperative of avoiding a possible British occupation, controlling trading and interoceanic navigation routes and maintaining surveillance of foreign maritime activities. However, from a local perspective, colonial agents viewed other economic benefits, namely, favoring the development of cattle production in Buenos Aires province, controlling and holding back from the south indigenous groups operating in the frontiers. The clearest references to this alternative of frontier policy design are found in the Patagonian commissioners' reports, who advocated for the convenience of maintaining southern posts and of fortifying strategic sites toward the interior of the territory in order to assure the control of foreign and indigenous movements and to prevent them from any possible alliance (Fig. 4.5).

The knowledge of the region, significantly enhanced by Jesuit production, resulted in the understanding of territorial organization and the identification of both suitable places for settlements and strategic sites for controlling the territory. Falkner and Cardiel's production was used during the implementation of the plan to people the area and in the exploration itineraries into the Patagonian interior, as well as a source of production of official cartography.

Juan de la Cruz Cano y Olmedilla (1776)[31] represents the territory of the Viceroyalty of the Rio de la Plata and possibly constitutes the cartographic image on which the occupation of the Patagonia coast was projected. Elaborated with materials from the Council of the Indies and maps that had been acquired by the author,[32] its references demonstrate the use of Jesuit maps—Quiroga,[33] Cardiel and Falkner[34], European maps—Bowen, Anson and Bellin, and other unidentified documents.[35] Its main contribution to the knowledge of the Pampean and north-Patagonian region is the depiction of the indigenous roads network, a key element to understanding the functioning of the cattle circuit during the last decades of the 18th century. It identifies two roads which cross the continent, from the Andes to the Pampas, which meet at *Guaminí*, near Sierra de la Ventana (*M. Casuhatí*). The northernmost one, called *Camino de los Pehuenches* or *Picunches*, runs toward the West up to the high valley of the Colorado River. The second one runs southwest to the Colorado River, connecting then to the Negro River at Choele-Choel, continuing farther along the northern coast of the river. A third road is also represented from the Buenos Aires frontier (the town of Pergamino) up to the Quinto River, with a reference that reads "unfrequented by Pampas' raids".

Cano y Olmedilla, *Mapa Geográfico de América Meridional* (1776)
http://bdh-rd.bne.es/viewer.vm?id=0000051497

The information obtained by Francisco de Viedma from "a cunning Indian, with experience in those vast lands" confirmed these routes:

> [...] the Choleechel was respectively pass of the Nations of Indians I have already mentioned [*Tehuelchú, Uhiliches, Puelches*, and other Nations living in *Huechuhuehuem*, and roundabouts of Valdivia and the Andes] but not of all of them; since there were other two roads along which they approached. One going into the salt flat that provided salt to Buenos Aires by cart and used also by the *Pehuenches*, and *Aucas*, whose Cacique was named *Guchulap*, and another one farther to the north, and not that far from the city of Córdoba, frequented by the *Ranquicheles* Nation.[36]

Knowing the existence of the routes, they could acknowledge the importance of the inter-hill valleys in Sierras de Tandil and Sierra de la Ventana, and the area of the salt flats in the Buenos Aires south, as provisions distribution and commercial interchange centers. Likewise, the strategic situation of *Cholchechel* or *El Paso* in order to control communication routes and trading became apparent.

The expedition led by Basilio Villarino in 1782 and 1783 to the source of the Negro River bolsters Falkner's information. It confirmed the pre-Andean piedmonts were inhabited by the *tehuelches, huiliches, pehuenches* and *aucaces* nations, while the coasts of the Negro and Limay Rivers—though not inhabited on a regular basis—were visited periodically as indicated by the evidence of shelters and roads. The pilot stated in his diary:

> These indigenous people have never been in our settlement of the Negro River. Although they say they know we were there, in order to go back to their lands, they have to walk 70 leagues to the west from our settlement, walking across the field from the Colorado to this river along the Chuelechel [...]. The places names, which other natives could not understand by reading Falkner, are identical as the ones in his log and coincide with him as regards the news (Villarino 1972: 1016–1017).

Among terrestrial explorations, Villarino's is best documented, since apart from the journey log, he provided a full sketch of the Negro-Limay waterway,[37] unique cartographic representation of the interior, resulting from the reconnaissance performed within the peopling plan. It is noteworthy that Antonio de Viedma's journey from San Julián to the Andean mountains in search of the sources of the Santa Cruz river left no cartographic register.

Villarino's contribution is two-fold. On the one hand, basing himself on the work of Falkner, he enhanced the image of the Negro River and confirmed data on the occupation of space. On the other hand, with information provided by native inhabitants, he broadened the view of the socioeconomic dynamics of the Pampas and northern Patagonia, describing the way in which groups operated to keep the vast interchanges network going—fueled by the movement of cattle—on indigenous lands and within the colonial market.

The purpose of Villarino's expedition was to assess the navigability conditions of the Negro River and to prove the possibility of communication with Valdivia through this river. The latter resulted to be a disturbing issue, since foreigners could use it.[38] However, the pilot's journey log shows an underlying interest toward gathering all the necessary information to design a colonization policy. This is evidenced by the thorough survey of lands suitable for agriculture and cattle breeding, remarks on natural resources, communication routes (roads, paths, passes, distances and walking time), settlements, areas of importance in the indigenous economy (watering places, pastures, hunting places and gathering areas) and strategic sites, particularly Choele-Choel and the confluence of rivers Negro and Neuquén. References to Falkner's account are quiet frequent, confirming the Jesuit's information in several passages of his journey log—mainly when referring to key areas if indigenous lands, commerce and toponymy—or else correcting it according to his own observations.

Information on cattle provision in the Pampas, its herding to the West and its sale to Hispanic-"*criollos*" in Chile confirmed the importance of commerce as an organizational axis of the economic space and the efficacy of indigenous territorial management. The same applied to the use of the Colorado salt flats and salt transport to the Andes and on the exchanged goods.

As regards the localization of the different groups, when comparing information provided by Villarino and that contained on Cardiel and Falkner's map, we observe significant variations. Having in mind the transformations that took place in the spatial dynamics, we wonder whether these differences could account for population movements due to their engagement in cattle trading during the middle decades of the 18th century. We will take into account one of the groups that appears on all maps—*Chulilauchet* on Cardiel's, *Chulilacunne* on Falkner's and related to Cacique *Chulilaquin* or *Chulilaquini* on Villarino's. We observe that while the two first ones locate it to the south of the Negro River, Villarino represented it in the pre-Andes region to the north of the Limay River, together with other *peguenches* and *aucaces* groups.

In the journey log, he provided accurate information on the territoriality and the movement entailed in taking part in cattle trading.

> *Chulilaquín* told me that his grandfather, main cacique of that land, was near the country of the apples [Neuquén province today], and that most of them were his relatives [...] his grandfather's *toldería* amounted to 80 to 100 huts: that *Huechum-lauquen* lagoon is behind a hill a native indicated to them, two leagues far from where they reached (Villarino 1972: 1112–1113).

> They say that all or almost all the indigenous people inhabiting the *sierras del Volcán* [Volcano hills, ancient name for Tandil] and Buenos Aires *pampas*, are from the west of this river [Negro], and that the reason why they spend so much time here is the big number of cattle and the possibility of feeding them. Some of them staying for two years, others longer, as it suits the (Villarino 1972: 1018–1019).

> [...] I found traces of those herding cattle to Valdivia before: but many more previous traces of a high number of cattle and horses, so many that, as I see it, there are more animals taken away from Buenos Aires than those that are consumed in that province (Villarino 1972: 1040).

The depiction of the Negro River basin on the *Mapa Geográfico* (1786),[39] produced by José Custodio Sáa y Faría[40] according to Villarino's sketch, enables the awareness of how important this river axis was within the socioeconomic organization of the Pampa space and the north of Patagonia. This is the first general map of the region, representing the basins of the Negro and Santa Cruz Rivers, according to the reconnaissance of the interior performed with native guides. Following Cardiel and Falkner's material and the maps, plans and reports produced within the peopling of the Patagonian coast plan, it represents the main river basins within the Atlantic slope, the relief, features on the coastline, communication routes, though most geographical elements appear to the north of the Colorado River. The course of this river is drawn from its mouth up to Choele-Choel, from where the indigenous route ran farther to the Andes. The high basin is not depicted, neither the basin of the Neuquén River. The unknown space is filled with the legend "Lands inhabited by several nations of barbarian indigenous peoples", which appears again in the interior of Patagonia. The outlining of the Santa Cruz River on a smaller scale, based on Antonio de Viedma's log, enables highlighting the importance of such river for communication with Chile in southern Patagonia.

Sáa y Faria, *Mapa Geográfico* (1786)
https://www.europeana.eu/portal/es/record/2022701/oai_rebae_mcu_es_176895.html?start=1&query=title%3A%22Patagonia+%5C%28Argentina%5C%29.+Mapas+generales.+1786%22&startPage=1&rows=24

Considering the variety of sources and the importance assigned to the accounts, journey logs and maps from imperial agents that collected spatial information from native inhabitants of the territory, Sáa y Farías' map could be thought as the synthesis of the state of regional geographic knowledge by the end of the 18th century. Nevertheless, the omission of essential information—available in the consulted corpus—to understand the territorial dynamics would demonstrate that there was an underlying intention in view of imminent interventions. The absence of information in previous maps could be due to lack of confirmation through European reconnaissance, mainly in the interior of the continent and to the south of the San Jorge Gulf (Porro Gutiérrez 1999: 139). Instead, the omission of indigenous routes known to the Spaniards as well as native toponyms, ethnonyms and names of chieftains that facilitated the localization of different ethnic groups, make us think of an information selection process conditioned by the purpose of the map within the context of the discussion on the region as a useful(less) place.

The *Mapa Geográfico* proves the significant advances in the knowledge of the interior and in the configuration of the Atlantic coast, anticipating José Espinoza y Tello and Felipe Bauzá's *Carta Esférica* and the depiction of the Santa Cruz River, which resulted from the reconnaissance performed by Fitz Roy. In spite of the fact that it did not have so much cartographic repercussion, it fulfilled its political purpose by enabling an intellectual appropriation of space and controlling strategies design. Sáa y Faría developed an argument that, responding to Kitchin-Falkner's map, gave credit to Spanish dominion on southern territories up to Magellan Strait and Malvinas Islands. Concurrently, it argued for the limited colonization of Patagonia, by representing it as a vast unknown territory controlled by independent indigenous nations.

4.5 Native Informants from the Southernmost Tip

Although the Magellan route had been a meeting space for Europeans and natives since the 16th century, the use of indigenous information in cartographic production is quite rare. Overall, contacts were brief and superficial, and though some accounts provide ethnographic descriptions, these result from observation and only exceptionally from communication with native inhabitants.

The Chart of Magellan Strait (1697),[41] drawn by Johan de la Guilbaudière—boatswain in a clandestine commercial expedition returning to Europe after operating at ports in the Pacific Ocean—was an early source that denotes transference of indigenous information, both verbal and possibly graphic one. The map was produced

Fig. 4.4 *Guilbaudière Carte du détroit de Magellan Indiens* (1697)

Fig. 4.5 Cano y Olmedilla *Mapa marítimo del Estrecho de Magallanes* (1769)

between 1684 and 1690, during a stay of almost one year on the western side of the strait, after the wreck on the coast known as *Joucqjoucq*, possibly Galiano Bay of Santa Isabel Island. During the months devoted to the construction of a new ship, exploratory routes were covered with the help of canoeing groups. The depiction of "few islands of large size with uniform coastlines, instead of an entangled sea as the western Magellan sector actually is" (Martinic B. 2016) could be construed as a topography stylish representation. That is, a feature of Amerindian cartography highlighted by Belyea (1996) and, thus, the way in which the *alacaluf* viewed their own territory. Another element that adds interest in this document is the fact that in it, there appear four ethnonyms identifying the distinct groups within an ethnic group.[42] Taking Guilbaudière's map as a source, La *Carte du Paraguay, du Chili, du Detroit de Magellan* de De L'Isle (1703) mentioned above, constitutes one of the first and few printed maps to include native toponyms, naming a channel that connects the Strait to the Southern Sea as *Jelouteché.*

Juan de la Cruz Cano y Olmedilla's *Mapa marítimo del Estrecho de Magallanes* (1769) amounts to another cartographic exception as a result of a profuse use of native toponyms in order to identify features of the Magellan coastline. Produced within a context of great concern on the maritime activities of rival powers, the map appeared to meet the needs of the Spanish Crown to uphold its rights on southern territories which were not left out of effective dominion, to reinforce control over maritime routes and guide the risky sailing through the interoceanic passage. This instrumental goal accounts for the conscientious selection of pertinent information, "based on several memoirs and journey accounts, either printed or manuscript, especially those of Sarmiento", as stated in the legend that appears next to the title.[43]

According to its references, Cano y Olmedilla's map uses documents from the expeditions led by Bartolomé and Gonzalo Nodal (1618–1619), John Narborough (1670–1671), Jorge Juan and Antonio Ulloa (1745) and John Byron (1764–1766). It draws special attention to Pedro Sarmiento de Gamboa's information indicating landmarks in his exploratory and foundation work and including indigenous toponyms registered during the 1580 reconnaissance. The log of this journey, commissioned by the Viceroy of Peru, revealed Sarmiento de Gamboa's interest in obtaining information from the natives in order to comply with the purposes of the expedition: to determine the route of the Strait, to identify suitable sites for future fortified settlements and spot possible English settlements.

The higher concentration of indigenous toponyms corresponds to the western area of Tierra del Fuego, seen by Europeans as the most adverse region due to its harsh weather conditions together with the particular conformation of the interior of the island. From the access to the Strait in the Pacific Ocean—*Puerto de Nuestra Señora de la Candelaria* or *Cuaviguilgua* and *Puerto de la Misericordia* or *Puchachailgua*—up to the middle sector, the map indicates the places where they had contact with the canoeing groups during the itinerary. It manifests the importance attributed to the native names given to ports, bays, inlets and coves as landmarks to reduce the risks of navigating through the intricate labyrinth of islands, islets, fjords and channels.[44] From the islands Santa Inés and Clarence up to the Atlantic Ocean, the kind of information varies: there are references to environmental and physiographic features, focusing mainly in natural resources—a penguin rookery, freshwater, good grasslands and forests—and identifying meeting places with natives—Bahía de la Gente, Bahía de los Gigantes, "Somewhere near here, Byron spotted the Giant Patagons in 1765". As regards the southern portion of Tierra del Fuego, the map presents brief descriptions of practices and space occupation—"Savages living on this side make fire when boats go by", "Dwelling of big indigenous people who cross the Strait by canoe made from tree bark".[45] Conceivably, the source of these references is 1670 Narborough's manuscript map,[46] which represents and describes natives in detail: "There are many people on both sides of the Strait; I've seen a lot of fires and smoke to the south of the entrance I named South Ulster. The natives cross the Straits by canoe, made from tree bark and there can be ten or twelve people on each; they stay on the southern side during summer" (Martinic and Moore 1982).

Some decades later, Robert Fitz Roy's observations confirm information on indigenous spatiality and agree on presenting the Magellan territory as an area of contact and commercial exchanges: "there appears to be every reason to conclude that the canoe-men on the southern side have frequent contact, and even a friendly one, with the horse-men in Patagonia". Later, he remarks that "in exchange for dogs, old horses, guanaco meat and old blankets, the former give pieces of iron pyrites (used to light a fire), their captives or their issue" (2013: 161–162).

Fitz Roy also noted the cooperation of members of canoeing groups in locating ports and resources of interest and described the mapping performed by a *chono* child, another noteworthy datum on indigenous cartography.

Niqueaccas knew the coast between 47° and the Strait of Magellan very well. After reaching a high hill, following a trip far from the view of the land, he pointed out the best ports and places to spot seals; and in some direction, and at a long distance (pointing at Seno Eyre, out of sight) he said there were plenty of seal furs.

On a quiet day, when nothing happened, he drew a map on the deck of the ship with chalk, covering from the mooring winch to the taffrail, and Mr. Low was sure he intended to say Seno Eyre, though by then, no white man had explored the waters on the western coast of Patagonia (2013: 78).

The captain highlights natives' "sagaciousness and vast local knowledge" and their "gift of recollection"[47] (2013: 178), notwithstanding that the cartography produced on the results of their exploratory work do not seem to comply with the instructions of the Admiralty, which expressly provided for the adoption of native toponyms, appreciating its descriptive strength.

Insignificant as it may seem, the intention of increasing the use of new names, with no precise meaning, tends to bias our geographical knowledge. The name assigned to a place by the first discoverer must be sacred for the common consensus of all nations. In new discoveries it would be more beneficial to assure that the name transmits some idea of the nature of the place; or *if inhabited, to adopt the native name*, rather than deploy the catalog of public characters or private friends (Fitz Roy 2013:50, the underlining is ours).

The first maps made on the data resulting from the surveys performed between 1826 and 1834[48] register few native names—five in the western area of the Magellan

Strait, taken from Cano y Olmedilla's map, and two in the core of the *Yagan* territory, incorporated by these explorations.[49] However, they had direct contact with the "*fueguinos*" (the Fuegian natives) during their stay in the region and Fitz Roy even took four youngsters to England and brought them back in the second trip.[50]

Maps published by the Hydrographic Office of the Admiralty were the source used—in Europe and also in the United States—to produce other general maps of Patagonia and of the Magellan region (Martinic 2011: 31). That is why modern cartography shows few references to the spatiality of native peoples in South America. *A chart of the straits of Magellan* (1776), an updated reissue of Cano y Olmedilla's *Mapa Marítimo* published by Jefferys, replaces Spanish for British names, though respecting some native names, later omitted on other Spanish and British maps. Thus, the 1789 Map of the Spanish Navy, synthesis of the hydrographic labor of Malaspina's expedition (1789–1794) and with plenty of toponyms on the coastline, includes no indigenous names and only presents the reference *Canal indio* in the West of the Magellan Strait.[51]

New States received—together with the cartographic production heritage—this "concealed agenda", accomplishing the emptying of indigenous elements in spaces addressed as "public lands", with an approach grounded on ideas of modernity and progress. During the second half of the 19th century, national toponymy became a key constituent of the process of territorial reinvention and the occlusion of indigenous geography. Most native names were limited to geographical features; that is, only associated to natural elements of the landscape (Lois 2014), and on the eve of the Conquest of the Desert, were intended to be kept due to their descriptive character as spatial reference (Ceballos 1986, Piana 1981).

The renaming strategy was particularly successful in southern insular territories, where symbolic appropriation preceded exploration and effective occupation. In the Magellan and Tierra del Fuego channels, with areas that remained almost unknown up to recent times, one of the few autochthonous names in use today in modern cartography exceptionally recalls indigenous territoriality in the name of a native informant. During a reconnaissance of western channels led by Pastor Thomas Bridges in the Anglican mission's schooner to Ushuaia, a Yagan man, originally from the region close to the shores of Peninsula Brecknock, informed about a channel unknown to the Europeans. "In our return journey from Punta Arenas to Ushuaia, we approached a mountain massif called Clarence Island. Acualisnan told us it was not an only island, since it was divided into two or three parts by a channel that runs across it. Guided by that skillful pilot, we went along an intricate itinerary. Acualisnan Channel can be seen today on the Admiralty maps, perpetuating his name" (Bridges 1978: 132). But even this finding of a new interconnection between the waters of the Magellan Strait and those of the Pacific Ocean, "it would take a long time to incorporate the news to cartography and, upon doing so, it was just doubtfully, as there was no serious survey" (Martinic 2011: 33).

4.6 Conclusions

The scrutiny of Jesuit maps, charts and accounts on the Southern borders of Chile and the Río de la Plata enables us to confirm that the missionary contribution was paramount to the construction process of the geographical knowledge of Patagonia and its cartographical representation. Integrating native knowledge within European referential systems, these materials broaden the understanding of the territorial dynamics, discover other ways of viewing and exploring the territory and constitute an essential source to comprehend indigenous territoriality in a time of deep socioeconomic transformations resulting from the inter-ethnic contacts.

The context in which it was produced favored the development of certain cartographical practices that privileged the spatial knowledge transmitted by native inhabitants. Natives' communication skills and the attention paid to the changing intra-ethnic relationships enabled the missionaries to carry out different procedures to obtain, register and validate the data. Jesuits could correct European maps used as a source for representations, descriptions and localizations of hydrographic networks, natural resources, settlements, routes and strategic sites to control the territory, by contrasting different versions in order to identify coincidences between different informants' testimonies and in different times, as well as the significant toponymy and collaborative mapping during itineraries guided by native experts.

Up to the last quarter of the 18th century, the state of European knowledge was limited to a few areas of the Atlantic coast and to the areas that were familiar to missionaries—hills in the southwest of Buenos Aires and the pre-Andes of Neuquén. Thus, references to the continental interior appearing on Jesuit maps could only derive from the access to the indigenous spatial knowledge. This was a specialized knowledge that guaranteed accurate lines of movements, control of routes and efficient use of resources. Such appropriation required the development of communicational and interpretational skills and the difficult task of harmonizing different spatial conceptions, experience-based and abstract space. Although these results are essential to the cartographic production, we note that the analysis tools and the nature of the available sources enable us to pose more queries than answers on the conditions that favored a particular receptivity in the "geographic translation" process.

Access to spatial information (distribution and location of physical elements) and territorial information (aspects related to socioeconomic organization) resulted in the production of territories as possible environments for missionary work. The Jesuit and illustrated cycle shared the same way of producing knowledge. Colonial field agents (missionaries, administrators representative of colonial authorities, explorers) constructed the data based on their exchanges with local informants, native inhabitants and Hispanic-"criollos" (deserters, prisoners, interpreters), privileging the qualified knowledge of chieftains. The strategies to obtain information were framed within hospitality and gift-exchange practices that outlined border relationships, mainly those of presents and parties and the compensation of services by means of European goods, highly appreciated by natives.

Considering the limited access to the scope of the evangelization project in the region and its interruption after the Order's expulsion from the Viceroyalty in 1767, we can observe that Jesuits were not the main beneficiaries of native knowledge. Instead, Jesuit cartography was utilized in exploratory trips, the design of occupation and territorial control policies and as a source of producing imperial cartography, by grasping the double image—European and indigenous—of southern territories and by providing information on the continental interior.

Colonial agents used these materials to build other meanings that, displacing the producers' intentions, served to develop and justify new projects. Appreciating this information corpus, Jesuits were considered the pioneers of American scientific cartography, though silencing "the people behind native toponyms on colonial maps or geographic discoveries promoted by well-known travelers" (Roller 2012).

Although the debate on the region as a useful(less) place re-created views of Patagonia as a barren land, conditioning the interpretation of space to categories determined by climatic conditions, the contribution of Jesuits as intermediaries of indigenous spatial knowledge allowed other representations and assessments of the territory. However, colonial policies did not contemplate the advance of colonization, limiting themselves to keeping a weak fort at the mouth of the Negro River and to promoting exploration. The latter helped them to obtain the necessary information to control the movement of groups that operated near the borders and to elaborate a cartographic speech that gave way to the intellectual appropriation of territories, not yet incorporated to the effective dominion.

Bearing in mind the important role of native knowledge, Cardiel's maps of Patagonia (1747 and 1751) and Falkner's (1772), Villarino's chart of the Negro River (1783) and that of the Magellan Strait by J. de la Guilbaudière (1697) could be considered "Amerindian maps". That is, based on European maps with information obtained from direct exchanges with natives and that communicate "someway" the sense of indigenous spatiality or the territorial conception of autochthonous peoples.

While Jesuit maps and some maps belonging to the illustrated cycle represents an inhabited territory and recognize the geographical knowledge of its inhabitants, the 19th century cartography shows a gradual loss of native names and references. The process will finish when national cartographies eliminate the last clues of access to the indigenous spatiality. This silencing strategy was particularly valuable when used on southern insular territories, where symbolic appropriation preceded state control.

Notes

1. According to Buisseret (2003: 16), the main function of maps is to transmit situational information: "The essence of a map is its ability to represent a local situation; maybe we should call it 'situational image', or even 'situational substitute'".
2. Belyea (1992, 1996), Ingold (2005), Thrower (2002) and Kok (2009) interestingly referred to indigenous cartography. Ingold affirms that indigenous maps—sketches and maps of the experience of movement of peoples—are a completely different cartographic expression when compared to Western maps,

and as expression of mapmaking aims mainly at guiding their displacement across the territory.

3. Data on the equipment used by Jesuits enable the assessment of knowledge and skills interplaying in exploring and cartographic labor. Cf. Furlong (1992).
4. The first results of this approach were referred to in Luiz (2006) and de Lasa and Luiz (2011).
5. We refer to the larger version of Ovalle's map, available at the repository John Carter Brown Map Collection. The two versions, both in Latin, propose the same contours, hydrographic network and relief, though differ in illustrations and toponymy. Cf. Hanisch (1976).
6. The official historian for the Society of Jesus remarked. "This map will be no contribution to Geographical Science, though it is worthwhile releasing it. It is absolute proof of the vagueness of ideas and the still prevailing uncertainty of mid-17th century as regards the interior of the American Continent. This ignorance became apparent in Ovalle in relation to Argentinean territory" (Furlong Cardiff 1936:31).
7. Ovalle acknowledged this information concealment on mines location or properties of herbs. As far as the City of Caesars is concerned, he did not mention having heard of it from the Mapuche people in Chile.
8. Ovalle might have based himself on Fray Gregorio de León's map of Chile, a lost manuscript that could have been a geographical chart or an account or both. In fact, he cited the work in six different occasions in which he refers to the gold in Chile, the indigenous people who are born with tails, the thickness of the trees on Chiloé, the ferocity of the land and its inhabitants' courage, the men covered in mud in the southern region and Colicheo's victory. Cf. Hanish 1976: 182, 257 and 258.
9. According to Furlong (1936), the editions of Ovalle and Rosales' works were contemporaneous though the latter remained unpublished until 1877.
10. *Mapa de la costa de los Patagones conforme al descubrimiento hecho de orden de SMC el año 1745 by A. P. Joseph Quiroga*, National Library, Madrid, Ms. 42/405.
11. *Diario del viaje que yo Dn Joachin de Olivares y Centeno, comandante de la Fragata de S.M. nombrada Sn Antonio, he hecho desde el Río de la Plata hasta el de Gallegos, al reconocimiento de la costa del sur* [...] (1745), National Library, Madrid, Ms. II/2778
12. G. De'Isle *Carte du Paraguay, du Chili, du Detroit de Magallan*, 1703. Repository: David Rumsey Map Collection, David Rumsey Map Center, Stanford Libraries. In another contemporary version of the map, reproduced by Martinic (1993: 83), the color of *tehuelches'* clothing reinforces the idea that the blankets were European.
13. J. B. D'Anville *Amerique Meridionale*, 1748. Repository: Hispanic Digital Library—National Library of Spain
14. For further information on Jesuit missions in Araucaria and the Pampas, refer to, among others, González de Agüero (1791), Furlong (1938), Nicoletti (2004),

Pinto Rodríguez (1988), Hernández Asensio (2001), 2003, Irurtia (2007), Martínez Martín (1994), Martins (2012), Nofri (2001), Asúa (2014).

15. As regards the border conflict on southern territories of the Spanish Empire, refer to, among others, Mandrini (1992, 1997), Palermo (1988, 1989), León Solís (1982, 1986, 1991), Pinto Rodríguez (1998), Ortelli (1996), Nacuzzi (2002), Luiz (2006).
16. Published by Outes (1940)
17. *Tierra de Magallanes con las Naciones que se han podido descubrir en viajes de Mar y Tierra desde el año 1745 hasta el de 1748*, National Library, Madrid, MR/ 42/362/1. Repository: Hispanic Digital Library—National Library of Spain.
18. *Mapa de Magallanes,* secondary author Santiago Palomares 1751, National Library, Madrid, MR/ 42/403. Repository: Hispanic Digital Library—National Library of Spain.
19. *A new map of the Southern Parts of America*, London, 1772, in Falkner (1974). Repository: John Carter Brown Library Map Collection
20. Letter written by Cardiel to the Governor and Captain General of Buenos Aires, dated on 11 August 1746, in Angelis 1969, T. II, pp. 548-558.
21. We support the cartographical analysis contained in the Letter and Account of Jesuit Missions in Paraguay Province, which Cardiel wrote to Father Pedro de Calatayud, on 20 December 1747, cited in Outes (1940); in *Diario del viaje y misión al río Sauce realizado en 1748* (Account of the voyage and mission to Sauce River carried out in 1748), in Publications of the Institute of Geography in the Philosophy School, A series, No 13, Buenos Aires, 1933 and the *Diario de su viaje desde Buenos Aires, al Volcán y Arroyo de la Ascensión* (Journal of the voyage from Buenos Aires to Volcán and Ascensión Stream) *(1752)*, in De Angelis, 1969, T. IV: 59-66.
22. Hernández Asensio (2003: 103) pointed out that "Spanish language was then the language of exchanges in the south of Buenos Aires and that, while a growing number of Indians-Pampeans, as well as Serranos, Aucas o Patagons—either reduced or free, knew how to speak Spanish quite well, there were very few Hispanic-"criollos" speaking any native language".
23. For further information on Amerindian peoples' conceptions of space, see Thrower (2002), Harley (2005).
24. According to Beylea (1996), communication is a kind of translation of spatial rather than verbal terms, requiring the creation of equivalent cartographic conventions to describe spatial features.
25. In his study on North American indigenous peoples' cartography, Thrower (2002) provides a similar reference when describing a manuscript map presented by an Iowa chief in a meeting held on Washington D.C. in 1837. It is a paper map, drawn in ink, possibly with a stick or a finger, sketching the vast area of the basin of the upper Mississippi and the Mississippi River. The author points out that "when comparing this outline to the same hydrography on modern maps it is clear that its reading is overall quite comprehensible".
26. "This lake is two-days from Valdivia, formed by several streams, springs and rivers originating in the Andes. Besides the river flowing to the east and south

forming part of the big river, maybe there exists another river flowing to the west, which probably communicates with the South Sea near Valdivia. However, I cannot uphold this since I was unable to examine it thoroughly" (Falkner 1974: 107; the underlining is our own). Villarino ratifies Falkner's ideas by referring to the lack of coincidence just on the distance from the lake to Valdivia, since "said account states two days while these natives assert it is four days", en De Angelis 1972, T. VIII B: 1017.

27. Some of the ethnonyms present spelling variations. The names Che che het and Pehuenches appear on both maps. On Cardiel's: Chulilauchet, Sehuagni, Muluches. On Falkner's: Chulilacunne, Sehoaucunne, Moluches.
28. T. Falkner A new map of the Southern Parts of America, 1772. Repository: John Carter Brown Library Map Collection
29. "This map is explained and some account is given from the country and its inhabitants in a pamphlet intended to be sold with the map".
30. The outlining of the river Camarones could indicate a compromise between Kitchin's decision to include it and Falkner's remark in the legend "River Camarones has been sketched under Mr. Bolton's authority, though I'm afraid it is imaginary, since it was not found in 1746 expedition".
31. J. Cano y Olmedilla's *Mapa Geográfico de América Meridional. Año 1776.* National Historical Archive of Spain, Maps, Plans and Drawings. Sigs. 676 and 738, 321 and 328. Repository: Hispanic Digital Library—National Library of Spain.
32. Among the materials from the Council of the Indies, Ferrand de Almeida (2009) identified more than 60 maps and plans, manuscript and printed, among which there were Spanish, French, English, Dutch and Portuguese maps.
33. It shows the legend "Bay of Sin Fondo or P. de S. Matías according to the maps of Quiroga's expedition".
34. Apart from the coincidences in toponyms and ethnonyms and in the distribution of elements, the legend that appears all along the course of the Negro River proves that the author knew Falkner's map: "Comó-Leuvú, gran Desaguadero, o R. de los Sauzes llamado en el Nuevo Mapa Ynglés Cusú, Leuvu o R. Negro". The use of certain ethnonyms shows the comparative use of Cardiel and Falkner's maps.
35. "*Laguna Coluguape desde cuya latitud principia el rio Gallegos según un manuscrito particular de estas regiones*"; "*Río Deseado sacado de un manuscrito moderno*"; "*Cabo Blanco situado según los manuscritos españoles*".
36. F. de Viedma to Marquis of Loreto, *Disertación* [...], Buenos Aires, May 1, 1784. National Historical Archive of Spain, State 2316 and General Archives of the Indies - Buenos Aires Sig. 99.
37. *Plano del Río Negro levantado por orden de D. Francisco de Viedma Comisario Superintendente de estos Establecimientos; por D. Basilio Villarino, Segundo Piloto de la Real Armada, en el reconocimiento que hizo de dicho río, desde su*

desagüe en el Océano hasta la Cordillera inmediata a Valdivia habiendo regresado al Establecimiento a los ocho meses de viaje. Reproducido en Cevallos 1986.

38. His observations evidence this priority through a detailed description of the river course (width, depth, speed, waterfalls), the daily register of distances and orientations and measures of latitudes.
39. J. C. Sáa y Faria *Mapa Geográfico que comprende todos los modernos descubrimientos de la Costa Patagónica y sus Puertos desde el Río de la Plata hasta el Puerto del Río Gallegos, junto al Cabo de las Vírgenes; la porción descubierta del Río Negro y caminos por la campaña desde Buenos Aires*, 1786. Repository: Catálogo Colectivo de la Red de Bibliotecas de los Archivos Estatales, Spain.
40. Sáa y Farías, Portuguese soldier and cartographer, took part in the outlining of the limit between Spanish and Portuguese American possessions under the provisions of 1750 Madrid Treaty and, since 1777, served for the authorities of the Viceroyalty of Rio de la Plata, participating in frontier issues.
41. J. de la Guilbaudière, *Carte du Detroit de Magellan* (1697). Repository: National Library of France - Department of Charts and Plans.
42. According to Martinic Beros (2011), taking into account the dialects used by surviving groups, ethnic designations—*Testeguiches, Lacquediches, Cadequiches* and *Hoquequediches*—would correspond to the local territories of the Kawéskar nation.
43. Cruz Cano y Olmedilla, *Mapa Marítimo del Estrecho de Magallanes*, 1769. Repository: National Digital Library of Chile and 1776 T. Jefferys' edition in David Rumsey Map Collection, David Rumsey Map Center, Stanford Libraries.
44. Toponyms registered in Sarmiento de Gamboa's account and on Cano y Olmedilla's map: *Xaultegua/Xualtegua, Exeaquil, Puchachailgua, Cuaviguilgua, Alguilgua, Capitloilgua, Caycayxixaisgua, Pelepelgua, Trinquichisgua.*
45. Such reference on navigation is provided by Falkner in his account when stating that chieftain *Tamu* had informed him "they use a buoy with which they sometimes cross the strait and, thus, they communicate with others from their own nation" (Falkner 1974:119)
46. J. Thornton, *A New Map of Magellan Narborough*, 1680. Repository: John Carter Brown Map Collection. See image in Chapter 3.
47. Conscious awareness, taking Ingold's concept (2005), or spatial memory.
48. Hydrographic Office of Great Britain, *The Strait of Magalhaens commonly called Magellan*, 1834. Repository: David Rumsey Historical Map Collection. Hydrographic Office, *The South-eastern part of Tierra del Fuego with Staaten Island, Cape Horn and Diego Ramirez Islands surveyed by Robert FitzRoy R.N. and the officers of H.M.S. Beagle 1830-1834, Staaten Island surveyed by Lieutenant E.N. Kendal in 1828*", 1841. Repository: Mystic Seaport Museum - G. W. Blunt White Library.
49. On the southern coastline of the Strait there appear *Periagua Island*, *Pt. Tinquichisgua*, *Gulf Xaultegua*, *Alquilqua B.* and *Puchachailgua* (located by

Cano y Olmedilla on the northern coastline), and in the region of Beagle Channel *Woollya* y *Tekenika Sound.*

50. They were a *Yagan* and three *alacalufes* baptized Boat Memory, York Minster (name of an island near Cape Horn), Fuegia Basket and Jimmy Button; the first one caught smallpox and died in the Naval Hospital and the others were vaccinated and lived during more than one year in Walthamstow, near London, at the expense of Fitz Roy (Bridges 1978).
51. *Carta esférica de las costas de América Meridional desde el paralelo 36 grados de latitud Sur hasta el Cabo de Hornos, levantada por orden del rey, en 1789, 90, 94 y 95 por varios oficiales de su real armada, presentada a su Majestad por mano del Excelentísimo Señor Don Juan de Langara, Secretario de Estado y del Despacho Universal de Marina. Año de 1789.* Repository: Madrid Naval Museum.

References

de Asúa M (2014) *Science in the Vanished Arcadia: Knowledge of Nature in the Jesuit Missions of Paraguay and Río de la Plata.* Leiden, Brill

Barcelos AHF (2007) Entre a cordilheira e o mar: exploração e evangelização jesuítica no Chile. Historia Unisinos 11(2):230–239

Bechis M (1997) Manipulaciones de Rosas en la Araucanía (1829-1831). Boletín de Historia y Geografía 13:49–68

Belyea B (1992) Amerindian maps: the explorer as translator. Journal of Historical Geography 18(3):267–277

Belyea B (1996) Inland Journey, Native Maps. Cartographica 33:2

Brañes MJ (2006) El chilidugú del padre Bernardo Havestadt. Introducción y selección. Onomázein 14:65–99

Bridges EL (1978) *El último confín de la tierra.* Buenos Aires, Marymar

Buisseret D (2003) *La revolución cartográfica en Europa, 1400-1800.* La representación de los nuevos mundos en la Europa del Renacimiento, Barcelona, Paidós

Burdick C (2014) Patagonian Cinnamon and Pepper: Blending Geography en Alonso de Ovalle's *Tabula Geographica Regni Chili* (1646). Imago Mundi 66(2):196–212

Cevallos, E. S. (1986) *La conquista de quince mil leguas. Estudio sobre la traslación de la frontera sur de la República al río Negro*, Buenos Aires, Hyspamérica

De Angelis, P. de (1969–1972) *Colección de obras y documentos relativos a la Historia Antigua y Moderna de las Provincias del Río de la Plata*, Buenos Aires, Plus Ultra

De Lasa, L. I de and Luiz, M. T. (2011) "Representaciones del espacio patagónico. Una interpretación de la cartografía jesuítica de los siglos XVII y XVIII", *Cuadernos de Historia* 35, Departamento de Ciencias Históricas—Universidad de Chile, pp. 7–33

Di Liscia MS, Prina AO (2002) Los saberes indígenas y la ciencia de la Ilustración. Revista Española de Antropología Americana 32:295–319

Falkner T (1974) *Descripción de la Patagonia y de las partes contiguas de la América del Sur.* Hachette, Buenos Aires

Fernández Bravo A (2004) Catálogo, colección y colonialismo interno: la *Descripción de la Patagonia* de Thomas Falkner (1774). Revista de Crítica Literaria Latinoamericana 30(60):229–249

Ferrand de Almeida A (2009) O Mapa Geográfico de América meridional, de Juan de la Cruz Cano y Olmedilla. Anais do Museu Paulista, São Paulo 17(2):79–89

Fitz Roy R (2013) *Viajes del "Adventure" y el "Beagle"*. Diario, Madrid, CSIC
Furlong Cardiff G (1936) *Cartografía jesuítica del Río de la Plata*, Buenos Aires, Facultad de Filosofía y Letras, Publicaciones del Instituto de Investigaciones Históricas Número LXXI
Furlong GSJ (1938) *Entre los Pampas de Buenos Aires*. Buenos Aires, Talleres Gráficos San Pablo
Furlong GSJ (1992) *Entre los Tehuelches de la Patagonia*. Buenos Aires, Teoría
González de Agüero P (1791) *Descripción e historial de la Provincia y Archipiélago de Chiloé en el Reino de Chile y Obispado de Concepción*, Madrid, Benito Cano, disponible en Memoria Chilena—Biblioteca Nacional de Chile, Digital
Hanisch W (1976) *El historiador Alonso de Ovalle*, Caracas. Facultad de Humanidades y Educación, Universidad Católica Andrés Bello, Instituto de Investigaciones Históricas
Harley JB (2005) *La nueva naturaleza de los mapas*. Ensayos sobre la historia de la cartografía, México, Fondo de Cultura Económica
Hernández Asensio R (2001) Fábulas y sueños de españoles…: el fracaso de la aventura jesuita en el sur de la provincia de Buenos Aires (1740–1753). Procesos, Revista Ecuatoriana de Historia, Quito 17:2–34
Hernández Asensio R (2003) "Caciques, jesuitas y chamanes en la frontera sur de Buenos Aires (1740–1753)", *Anuario de Estudios Americanos*, t. LX-1, Sevilla, pp. 77-108
Ingold T (2005) The *perception of the environment: in livelihood, dwelling and skill*. Routledge, London
Irurtia MP (2007) Intercambio, novedad y estrategias: las misiones jesuíticas del sur desde la perspectiva indígena. Avá 11:135–167
Jiménez JF, Alioto SL (2016) "Recorredores de mundos: viajeros nativos en las pampas y Araucanía (siglos XVIII y XIX). Revista colombiana de antropología 52(1):245–270
Justo, M. S. (2015) "Revisitando la descripción de la Patagonia del padre Thomas Falkner. Modelos retóricos y escritura jesuita", *Atek Na Historia*, 5
Kitchin, R. and Perkins, C. (2009) "Thinking about maps", in: Dodge, M., Kitchin, R. and Perkins, C., *Rethinking Maps, London, Taylor & Francis Ltd*
Kok G (2009) Vestigios indígenas en la cartografía del sertão de la América portuguesa. Anais do Museu Paulista 17(2):91–109
León Solís L (1982) La corona española y las guerras intestinas entre indígenas de Araucanía, Patagonia y Las Pampas, 1760-1806. Nueva Historia, London 2(5):31–67
León Solís L (1986) Las invasiones indígenas contra las localidades fronterizas de Buenos Aires, Cuyo y Chile, 1700-1800. Boletín Americanista, Barcelona, XXVIII 36:75–104
León Solís L (1991) *Maloqueros y conchavadores en Araucanía y las Pampas, 1700–1800*, Temuco (Chile), Universidad de la Frontera
Livon-Grosman E (2003) *Geografías imaginarias*. El relato del viaje y la construcción del espacio patagónico, Rosario, Beatriz Viterbo
Lois C (2014) *Mapas para la nación: episodios en la historia de la cartografía argentina*. Buenos Aires, Biblos
Luiz, M. T. (2006) *Relaciones fronterizas en Patagonia. La convivencia hispano-indígena a fines del período colonial*, Ushuaia, Universidad Nacional de la Patagonia San Juan Bosco—Asociación HANIS
Luiz, M. T. and Schillat M. (1997) *La frontera austral. Tierra del Fuego 1520–1920*, Cádiz, Servicio de Publicaciones de la Universidad de Cádiz
Lynch K (1984) *La imagen de la ciudad*. Gustavo Gili, Barcelona
Mandrini, R. (1992) "Indios y fronteras en el área pampeana (Siglos XVI y XIX). Balance y perspectivas", Anuario del IEHS nº 7, Tandil, Universidad Nacional del Centro de la Provincia de Buenos Aires, pp. 59–72
Mandrini R (1997) "Las fronteras y la sociedad indígena en el ámbito pampeano", *Anuario del IEHS* nº 12. Universidad Nacional del Centro de la Provincia de Buenos Aires, Tandil, pp 23–34
Martínez Martín C (1994) Las reducciones de los Pampas (1740-1753): aportaciones etnogeográficas al sur de Buenos Aires. Revista Complutense de Historia de América, Madrid 20:145–167

Martinic B, M, Moore, D. M. (1982) Las exploraciones inglesas en el Estrecho de Magallanes 1670–1671. El mapa manuscrito de John Narborough", *Anales del Instituto de la Patagonia.* Punta Arenas (Chile) 13:7–20

Martinic M (1999) Cartografía magallánica 1523–1945. Universidad de Magallanes, Punta Arenas

Martinic Beros, M. (2011) *El occidente fueguino todavía una incógnita,* Punta Arenas (Chile), La Prensa Austral

Martinic Beros M (2016) Bucaneros en el estrecho de Magallanes durante la segunda mitad del siglo XVII, nuevos antecedentes. Magallania (Chile) 44(1):5–14

Martins, M. C. B. (2012) "As missoes de pampas e serranos: uma experiencia de frontera na pampa argentina (Século XVIII)", *Clio—Revista de Pesquisa Historica,* 1 (30): 1–17

Martins MCB (2017) Cultura escrita e projetos coloniais: "A Descrição da Patagonia" de Thomas Falkner. Revista Maracanan 16:34–51

Mörner M (1985) *Actividades Políticas y Económicas de los Jesuitas en el Río de la Plata.* Buenos Aires, Hyspamérica

Nacuzzi, L. R. (comp.) (2002) *Funcionarios, diplomáticos, guerreros. Miradas hacia el otro en las fronteras de Pampa y Patagonia (Siglos XVIII y XIX),* Buenos Aires, Sociedad Argentina de Antropología

Nicoletti, M. A. (2004) "Los jesuitas en el Nahuel Huapi: aproximaciones a una breve misión en la Patagonia (1669–1717)", *Memoria Americana. Cuadernos de Etnohistoria* Nº 12, Instituto de Ciencias Antropológicas, Facultad de Filosofía y Letras, Universidad de Buenos Aires

Nofri MC (2001) Actividad misional y resistencia indígena en las reducciones jesuitas de Pampas y Serranos (1740-1753). VIII Jornadas Inter-Escuelas y Departamentos de Historia, Salta

Ortelli S (1996) "La araucanización de las pampas: ¿realidad histórica o construcción de los etnólogos?", Anuario del IEHS, no 11. Universidad del Centro de la Provincia de Buenos Aires, Tandil, pp 203–225

Outes FF (1940) *Carta inédita de la extremidad austral de América construida por el P. José Cardiel S. J. en 1747,* Buenos Aires, Universidad de Buenos Aires, Museo Etnográfico, Publicaciones B, nº 1

Palermo MA (1988) "La innovación agropecuaria entre los indígenas pampeano-patagónicos. Génesis y procesos", *Anuario del IEHS,* nº3, Tandil, Universidad del Centro de la Provincia de Buenos Aires, pp. 45–90

Palermo MA (1989) Indígenas en el mercado colonial. Buenos Aires. Ciencia Hoy 1(4):22–26

Piana EL (1981) *Toponimia y arqueología del siglo XIX en La Pampa.* Buenos Aires, Eudeba

Pinto Rodríguez J (1988) "Frontera, misiones y misioneros en Chile: La Araucanía 1600-1900", *Misioneros en la Araucanía (1600-1900): un capítulo en la historia fronteriza de Chile.* Universidad de la Frontera, Temuco

Pinto Rodríguez, J. (1998) "Araucanía y Pampas. Una economía fronteriza en el siglo XVIII", *Boletín de Historia y Geografía,* nº 14, Santiago de Chile, Universidad Católica Blas Cañas, pp. 197–221

Porro Gutiérrez JM (1999) *Introducción a la cartografía histórica americana.* Universidad de Valladolid, Valladolid

Quintero, S. (2000) "Pensar los mapas. Notas para una discusión sobre los usos de la cartografía en la investigación social", in: Escolar, C. (comp.), *Topografías de la investigación. Métodos, espacios y prácticas profesionales,* Buenos Aires, Eudeba, pp. 187–217

Roller HF (2012) River Guides, Geographical Informants and Colonial Field Agents in the Portuguese Amazon. Colonial Latin American Review 21(1):101–126

Sáa y Faría, J. C. (1972) Respuesta del Brigadier don José Custodio de Sáa y Faria a S.E. sobre el reconocimiento y diario de Villarino [1783], en De Angelis, P., *Colección de obras y documentos,* t. VIII b, Buenos Aires, Plus Ultra, pp. 1147–1150

Sponberg Pedley M (2005) *The Commerce of Cartography: Marking and Marketing Maps in Eighteenth-Century France and England.* The University of Chicago Press, Chicago

Torre Revello J (1938) *Mapas y planos referentes al Virreinato del Plata conservados en el Archivo General de Simancas*, Buenos Aires, Facultad de Filosofía y Letras, Publicaciones del Instituto de Investigaciones Históricas, LXXXIII

Thrower NJW (2002) *Mapas y civilización*. Historia de la cartografía en su contexto cultural y social, Barcelona, Ediciones del Serbal

Udías A (2015) *Jesuit Contribution to Science: A History*. Springer, London

del Valle I (2009) *Escribiendo desde los márgenes*. Colonialismo y jesuitas en el siglo XVIII, México, Siglo XXI

Villalobos RS (1989) *Los pehuenches en la vida fronteriza*. Ediciones Universidad Católica de Chile, Santiago de Chile

Villarino B (1972) *Diario del Piloto de la Real Armada D. Basilio Villarino del reconocimiento que hizo del Río Negro en la Costa Oriental de Patagonia el año 1782*, in De Angelis, P., *Colección de obras y documentos*, t. VIII B, Buenos Aires, Plus Ultra, pp 967–113

Chapter 5
Southern Patagonia, Tierra Del Fuego, the South Atlantic and The Antarctic Lands Within Global Strategies

Abstract In this chapter, we approach the integration of the southern region into the capitalist system and the state territorial designs in the 19th century. The study of cartography, based on other documental sources, enables the recognition of spatial representations with underlying ideologies and interests of different actors. Similarly, we consider the coexistence of territorialities with several origins and their own contents and logics, which define the specific rationality of a space of growing geopolitical and economic importance. We identify a cartographical narrative that, together with the imperial ideology of its authors and/or promoters, served the purposes of the British expansion propaganda in the southern hemisphere. Apart from state territorialities—which are shaped on indigenous territories and overlap with each other, triggering sovereignty disputes—we note the development of private projects that influence the valorization of the southern tip of the continent and account for the territorial dynamics throughout the last decades of the 19th century.

Keywords Representations of space · Territorialities · Cartographic strategies · Patagonia · Tierra Del Fuego · The South Atlantic and the Antarctic · 19th century

5.1 Introduction

During the 18th century, while expectation of discovering new inhabitable lands in the southern hemisphere weakened, European expansion and commercial plans started to consider the southwestern Atlantic Ocean. Mainly in the last half of that century, within a context of a renewed interest in the control over navigation interoceanic routes and the exploitation of abundant faunal marine resources, the southern territories in America achieved their own entity in European cartography. However, there were still representations that relate it to polar lands.

Maps show southern Patagonia, Tierra del Fuego, the Malvinas and—since the first decades of the 19th century—the sub-Antarctic islands and the coasts known as the Antarctic Peninsula as part of a South Atlantic region. Partially explored and not accurately outlined, this space appears as a region in the geopolitical sense, that is, an area of state interest on which they plan actions and anticipate effects.

L. I. de Lasa and M. T. Luiz, *The Southernmost End of South America Through Cartography*, The Latin American Studies Book Series,
https://doi.org/10.1007/978-3-030-65879-3_5

Cartography takes part in the development of spatial conceptions and territorial images, that is, those aimed at legitimating beliefs, interests, projects and interventions that condition the ways of thinking of and valorizing space. In the mid and long term, they set geographic imaginaries and territorial identities. This construction, in which scientific discourse and journeys literature play their part, constitutes the symbolic dimension of the territorial appropriation.

Cartographical analysis enables the recognition of representations of southern territories with underlying ideologies and interests of different actors. While Europeans depicted Patagonia as a savage space, under no state dominion, cartographical omissions paved the way to their incorporation into the new national states, promoting the idea of a "desert" that could be transformed from the capitalist modes of production. On the other hand, representing the South Atlantic Ocean and the adjoining area of the Antarctic continent as an area of British influence or dominion was part of a cartographic discourse that, together with political and commercial motives and the imperial ideology of its authors and promoters, served the purpose of the territorial expansion propaganda and the construction of a British territoriality in the South Atlantic Ocean.

Apart from the overlapping of national and imperial frontiers in the South Atlantic Ocean, and the national states territorialities shaped on indigenous territories and overlapping with each other, thus generating frontier controversies, we observe the active role of other actors who influence the valorization of the territory and account for its functioning. On the one hand, religious institutions define a transnational missionary territory and, on the other hand, the expansion ventures of business people related to sheep livestock, transport and finance, favored by Argentinean and Chilean policies. This leads to the incorporation of the territories of Magellan, Santa Cruz and Tierra del Fuego into the capitalist system during the last decades of the 19th century.

5.2 From Inhospitable Land to "Nobody's Land"

Cartography scrutiny made during the first decades of the 19th century uncovered continuities and changes upon the representations of Patagonia. Cartographical legends—"unknown land", "country inhabited by nomads and savages", "desert land"—keep on recreating the stereotyped and ambivalent images typical of how the western imaginary thought of any place 'there'" (Staszak 2012). While bolstering the idea of a forbidding region of doubtful utility, they describe an unexplored (unexploited) and attractive space. If the idea of an unknown territory controlled by natives had justified the limited Hispanic-"criollo" peopling toward the southern frontiers of the empire, the end of the Spanish era provided enough grounds for expansionist projects of different powers and new independent republics.

All European states colonizing America turned to the discovery doctrine in order to legitimate the conquest of indigenous territories (Miller et al. 2010). The Crown of Castile sought to assure its exclusive rights in the New World through the papal donation (1493). However, after the reaction of John II of Portugal, the Treaty of Tordesillas (1494) set up a line of demarcation of jurisdictions that, subsequently, accounted for Portuguese sovereignty over Brazil (Bennassar 1987). Even though the *eximie devotionis* bull had granted not only the right to conquer but also to enslave the inhabitants of Antilles, the Crown was never convinced of making use of the native workforce. Therefore, by the mid-16th century, after the doctrinal debate on the "legal titles" by conquest and at the climax of the debate on the human, social and legal condition of American natives, papal bulls were deemed as mere territorial concessions with no effect at all on their self-determination (Levaggi 2000; Pagden 1988). From this point of view, the violent conquest carried out during the first decades and legitimated by the first "Requirement" was replaced by a peaceful conquest policy.[1]

Indigenous resistance gradually outlined the empire frontiers and the limits of colonial dominion. The adoption of a special law, based on the Law of Nations, and of a policy supported by trading and diplomacy, enabled the regulation of the treatment of non-subjugated peoples that, by mid-18th century, controlled half the territory now known as Hispanic America (Levaggi 2000; Weber 1998).

Bourbon administrators privileged the settling of agreements with natives in order to assure the frontiers security and prevent potential interventions of rival powers. This strategy favored the development of a series of transactions of political and economic nature that strengthened the autonomy of the societies in contact with each other. The entering into treaties, written or verbal, with social and legal scope, implied the recognition of natives as constituents of a nation as a legal entity, who exerted effective control over their territories (Levaggi 2000: 563). Although they could voluntarily accept vassalage and Spanish sovereignty, as observed in Parliament's acts and wording of treaties, it was somehow contradicting the strength with which they could actually negotiate while being the "king's vassals" (Lacoste 2003: 146). The recognition of chieftains' power and authority, reinforced by the granting of titles and political posts (chieftains ambassadors, chieftains governors) and the honors and gifts policy[2] that on occasions turned the Hispanic-"criollos" into "tributaries" of the natives (Luiz 2006) contributed to a political autonomy within a frontier context of increasing socioeconomic interdependency. Although technologies of colonial powers "tended to create discrete indigenous entities in order to absorb them later within the State" (Boccara 2003: 85), the Bourbons were not the beneficiaries of said policy but the republican governments were.

The agreements system proved to be effective during the political transition of the Revolution, when Araucanian and Pehuenches chieftains supported the Royalists. Such an alliance mobilized a powerful guerrilla at the south Andean frontiers between 1818 and 1832 (Manara 2005). Later, while Chilean authorities resumed the colonial indigenous policy based on evangelization and the parliament, negotiations with different partialities during Juan M. de Rosas' government in Buenos Aires (1829–1852) favored the so-called "peaceful business of Indians". The frontier policy of Buenos Aires involved allied independent groups who agreed to help them to keep peace at the frontier in exchange for cattle and other goods, as well as friendly natives, tribes that had settled on territories controlled by the provincial state and had gradually lost political autonomy (Quijada 2002; Ratto 2003). Although the status of the free groups suffered no significant change and was still regulated by the Law of Nations until the military campaign (1879–1884), the "criollo" dominance was gradually imposed through subtle mechanisms during the 1830s and the 1840s, namely the natives' integration into the productive and administrative system and the appropriation of indigenous territories by the State and by private individuals (Boccara and Seguel-Boccara 1999; Levaggi 2000).

The cartographic discourse of the late colonial period, expressed through J. C. Cano y Olmedilla's maps (1779)[3] and M. Lastarria (1804),[4] was able to coordinate the symbolic possession with the actual occupation and current possession, by incorporating territories over which there was no effective control into the empire. However, cartography used different techniques—coloring, toponyms and ethnonyms, legends and other inscriptions—to distinguish colonial and indigenous jurisdictions.

The integration of Araucania and the Pampas into the Governorate of Chile and the Viceroyalty of the Río de la Plata, respectively, could have been justified by the strong links between indigenous territories and the colonial centers on both sides of the Andes due to trading exchanges or else due to the treaties and alliances scheme between Hispanic-"criollos" and natives.[5] However, the dominion on southern territories was kept as a formal issue. The gradient of the Pacific Ocean, to the south of the Bio Bio River (37° south latitude), settlements scarcely amounted to three: Valdivia and Osorno—founded by the mid-16th century and abandoned after an attack from indigenous tribes in 1604 and repopulated in 1647 and 1793, respectively—and Chiloé Island colonized from the foundation of Castro in 1567. To the east of the Andes, the actual frontier was the line of forts constructed along parallel 36° south latitude, from Cuyo to the south of the Buenos Aires province, and to the south, only Carmen de Patagones survived, built on the fort established at the mouth of the Río Negro in 1779, and the colony and prison at Puerto Soledad on the Malvinas Islands, dependent on Buenos Aires since 1767 until the islands were abandoned in 1811 (Fig. 5.1).

Upon the territorial dismemberment of the Iberian Empires, during the formative stage of new state frontiers, European cartography gives hints at the intent of legitimating projects of various types, and the political and economic interests of various actors.

England and France were instrumental in the development of the doctrine of discovery, adding the concepts of "actual occupancy and current possession" as requirements to assert territorial claims and elaborated specific grounds to legitimize the occupancy of indigenous territories. The principle of "*terra nullius*" or "vacant land" "dictated that lands which were not possessed by any person or nation, or which were occupied by non-Europeans but were not being used in a manner European legal systems approved, were considered to be waste and vacant" (Miller et al. 2010: 848–849).

Cartography was an effective instrument in this legal-political construction that—from an ethnocentric assessment of cultural behaviors set a hierarchy of human groups—established the superiority of the Occident and defined the Others' possibilities and impossibilities of reaching civilization. The first thematic maps of the

Fig. 5.1 Lastarria *Mapa Geográfico de América Meridional* y *Carta Corográfica del Virreinato* (1804)

world that include demography forms of government and religion show the ideological compromise and political intent of the agencies. Those elaborated by J. Wyld (1815) and W. Woodbridge (1821 and 1827)[6] visually reinforce the central role of Europe and the United States and present a distribution of settlements based not only on space, since—as of the mid-18th century—the classification of civilization status indicated positions as regards a temporal scale, acknowledging cultural differences as a result of the rhythms of historical development (Pagden 1988: 23).

Maps reproduce a classification-qualification scheme, in which the term "savage" and "barbarian" are not yet vaguely used to refer to the peoples that were not Christian or culturally sophisticated enough (Pagden 1988: 46). Such terms adopt a conceptual accuracy when used to refer to settlements with less civilizing potential and impose a differentiation between nations and territories that resulted functional to the colonial projects and the goals of the new national states.

The scientific discourse of the first half of the 19th century shows the transition from the Enlightenment concept of the "noble savage" of the travelers of the 18th century to the belief that the natives who were not subjugated were savages and inherently insubordinate. Both Antonio de Córdoba and Alejandro Malaspina had described the inhabitants of the maritime coasts of the insular tip of America as savages, on the grounds of their human degeneration resulting from their isolation in regions deemed by Europeans as the most barren. Both of them had also distinguished Yaghans and Alacalufes "savages" and the Tehuelches, Onas and Chonos "barbarians", who were able to adopt western cultural habits (Navarro Floria 2003). However, during the first half of the 19th century, "the Enlightenment trust in indigenous people's *civilization*, although potential, led to the acceptance of an *inevitable* submission" (Navarro Floria 2005) since, as A. D'Orbigny[7] states, insubordination obeyed the immutable laws of nature.

The condition of freedom, associated to paganism and political fragmentation, was essential for subscribing to this savage theory and for applying the *terra nullius* principle. However, on occasions, the corpus of science makes it possible to anticipate the deficiency of the argument used to legitimate the advance on territories neither subject to the state control nor appropriated according to capitalist guidelines. In his *Geografía Universal*, W. Woodbridge describes the forms of authority in the independent Araucanian tribes and, after noting the existence of hereditary leadership and power concentration in times of war, concludes that "their administrations are as regular as those of civilized nations" (Fig. 5.2).[8]

In turn, the well-known geographer Adriano Balbi provides a classification of societies in the New World on the basis of language and political organization, two categories that according to the Greeks and later the Romans defined the existence of a civil community.[9] Bearing in mind the languages of the main nations, he distinguishes two big sets, the American peoples or indigenous and those of foreign origin. He also acknowledges the difficulty in "determining the number of the many indigenous civilization focuses and outlining their limits" (Balbi 1848: 222). Regarding the forms of government of the recognizable political entities, he identifies an "independent indigenous America", which comprises "a big number of small states formed by the nations that, although scattered along vast spaces that European powers and the new States of America consider as part of their respective territories, still kept their independence" (Balbi 1848: 399). The geographer points out that the nations gathered in "the powerful Araucanian Confederation", "have a regular government", "a combination of aristocracy and democracy", and he compares it to the structures of European powers:

Fig. 5.2 Woodbridge *Moral and Political Chart of the World* (1827)

> The government of this country provides the most significant resemblance to the military aristocracy of the dukes, counts and marquis of the north in the Old World. However, they existed prior to the arrival of the Spanish at that distant region in the New World. The Araucanians are the most civilized independent indigenous nation in America. Also, they seem to be the first people in that continent who-having a good number of horses-got used to horse riding and formed cavalry units. Since 1563, they already had many squadrons in their military formations. (Balbi 1848: 366).

However, such a perception of the high degree of socio-political organization of the Mapuches was not enough to change the generic conceptualization of the natives as "nomadic savages" (Quijada 1999). Brue (1843)[10] cartographically expresses this conception of the existence of a cultural hierarchy by integrating the nations considered close to civilization, the Araucanians and related groups settled to the east of the Andes—Pehuenches, Puelches, Ranqueles and Pampas—into Chile and the Provinces of the Río de la Plata. Likewise, he represents Patagonia to the south of Río Negro as a vast independent territory inhabited by nomadic peoples, the Tehuelets or Patagonians.

During the first decades of the 19th century, there were maps that still depicted colonial delimitations while others already represented the territorial fragmentation supervening the imperial crisis. As Lois (2015) remarked, it expressed the different ways of thinking, understanding, imagining and anticipating the political configurations of the region. Most of the first ones represent southern territories under Spanish formal dominion whereas most of the latter distinguish two state territorial units—Chile and the United Provinces or La Plata—from a vast bi-oceanic entity, identified with the names of Patagonia, Magellan Lands, *Chica*, New Chile or Modern Chile, and comprised the indigenous territories on both sides of the Andes and Tierra del

Fuego. The Araucania is generally deemed as part of Chile and, to the east of the Andes, the indigenous border in the northern Patagonia and the Pampas is located between parallel 36° south latitude and the Negro River. Some maps, adopting the criterion of effective occupation, place it over the line of forts that extended from the south of Mendoza to the coast of Buenos Aires (Wilkinson 1806).[11] Others, based on Cano y Olmedilla's design, incorporate indigenous territories within the cattle commercial circuit up to the Colorado and Negro Rivers into the Río de la Plata jurisdiction (Cary 1807).[12]

L. Delarochette's map (1811), a different version of W. Faden's map of South America (1807), shows an image prior to the territorial disintegration, by representing the formal boundaries of the Spanish Empire by means of coloring. It incorporates Patagonia, Tierra del Fuego and the Malvinas Islands into the colonial space and sketches two internal delimitations. The first one over the Snowy Andes Mountain Range, designating the limit between the Captaincy General of Chile and the Viceroyalty of the Río de la Plata jurisdictions, and the second one a straight line that indicates the actual indigenous border, from the area occupied by the Jesuit missions to the south of the Salado River up to the Andes (Figs. 5.3 and 5.4).[13]

The governments in Buenos Aires and Santiago claimed the jurisdictional continuity up to the southern tip of the continent. Concurrently, European and North-American cartography settled an image of the southern frontier of Chile at 44° south latitude and that of the jurisdiction of the United Provinces at the 41° south latitude, up to the Negro River. Indigenous ethnonyms appearing within these limits and beyond them as well—Araucanians, Aucas, Ranqueles, Pampas, Puelches, Huilliches, Pehuenches and Tehuelches—indicate the areas controlled by different independent tribes. Naturalist A. D' Orbigny (1999: 529–530), who stayed at Carmen de Patagones for eight months in 1829, pointed out the arbitrary adoption of this delimitation in French and foreign maps, observing that Argentinian dominion reached the San José Peninsula (Valdés Peninsula) on the Atlantic coast. The indigenous territory, in turn, spread in the interior up to the Colorado River and Sierra de la Ventana.

The successive editions of A. Brue's map of southern America depict the shifting of the Pampean-North Patagonian frontier from the line of forts along parallel 36° south latitude (Brue 1816)[14] to the Colorado River (Brue 1826)[15] and, since the 1830s, its consolidation at the Negro River (Brue 1839).[16] Brue's map of 1843 mentioned above shows fewer native toponyms to the south. It only highlights the ethnonym "Araucanians" and keeps a couple of indigenous names in the southern tip (*Xautalgua* and *Tekenika*). Resulting from Parker King and Fitz Roy's reconnaissance, English names prevail in Patagonia and Tierra del Fuego, while most of the toponyms are Spanish on the coasts of the Atlantic Ocean. This map also outlines the indigenous border to the south of the Bio Bio River, all along the way between Concepción and Calbuco, which separates the coastal area from the area by the Andes, controlled by the Mapuches (Fig. 5.5).

One of the features of the maps of South America and regional maps of the first half of the 19th century is the dubious correspondence between the several devices used to indicate the different status of dominion, namely, coloring to differentiate geographical and political units, toponyms, ethnonyms and legends referring to discovery,

Fig. 5.3 Wilkinson *South America* (1806)

exploration and occupation antecedents. Although the inaccuracy of the boundaries of the administrative colonial units and the post-independence changing political panorama contributed to the confusion, map producers and promoters' intentionalities cannot be ignored in the context in which territorial and commercial expansion expectations are renewed. Cartographical representation of the southern territories, the South Atlantic Ocean archipelago and the known Antarctic Peninsula coasts offers a glimpse of economic interests and geopolitical imaginaries. For example, L. Delarochette (1811) represents the Malvinas Islands as a Spanish domain, but he uses the toponym Falkland together with "Hawkins's Maideland"/"Isles Malouines

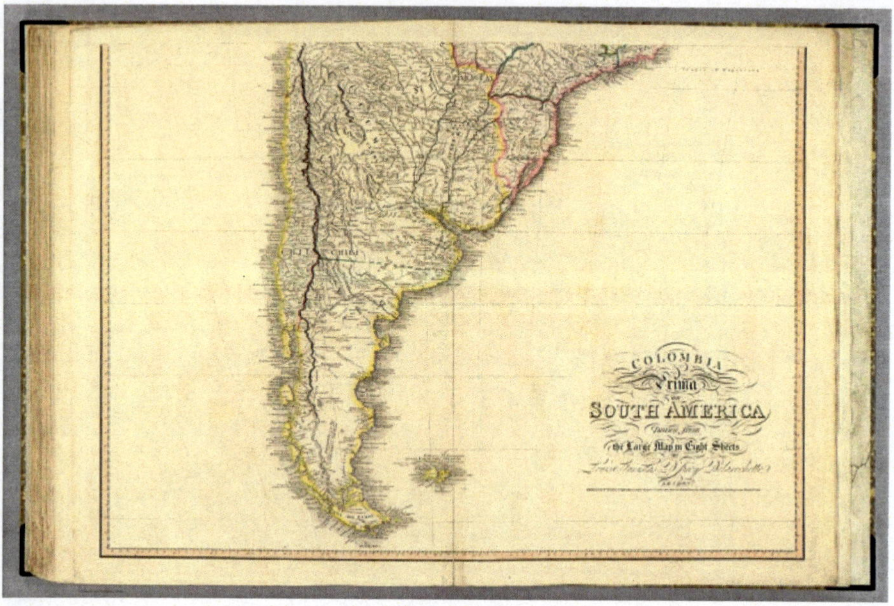

Fig. 5.4 Delarochette *Colombia Prima or South America* (1811)

of the French 1770", and the maps of C. Malte Brun (1837) and Tanner (1836) add this archipelago to *La Plata* through coloring, although keeping the French and English toponymy.[17] A. Brue's maps (1821, 1826, 1839) include Patagonia, Tierra del Fuego, the Malvinas Islands, the sub-Antarctic islands—Georgias, Orkney, Shetland and Sandwich—and the Antarctic lands among the territories referred as "not conquered Country" and "independent Country".[18] These names, together with other names that describe territories less known and inhabited—"Unknown Country" and "Uninhabited Country" (Cary 1807, Thomson 1814), "Desert Land" (Bradford 1838, Greenleaf 1840)—reinforce the representation of a vacant space, not part of any state sovereignty.

North American cartographical publications provide different territorial designs, although as Lois remarked (2015: 301), it is likely that the atlases "in general, predictably empathize with the emancipatory tendencies due to their own recent history have shown themselves particularly sensitive to revolutionary processes that affected Latin America". H. Tanner's maps, included in the most popular atlas in the country between 1834 and 1860, represent the southern territories as spaces of formal dominion of new republics and set the limit in the Andes, thought of as a natural frontier.[19] T. Bradford, another cartographer born in Boston, claims for the differentiation of the three territorial entities appearing in European maps: Chile includes Chiloé Island and the territory inhabited by the Araucanians—"a powerful independent nation of about 80,000 souls"—the United Provinces extend up to the Negro River and the Patagonia, between the 36° and 56° south latitude, appears as territory "inhabited by independent native tribes and not claimed by any civilized

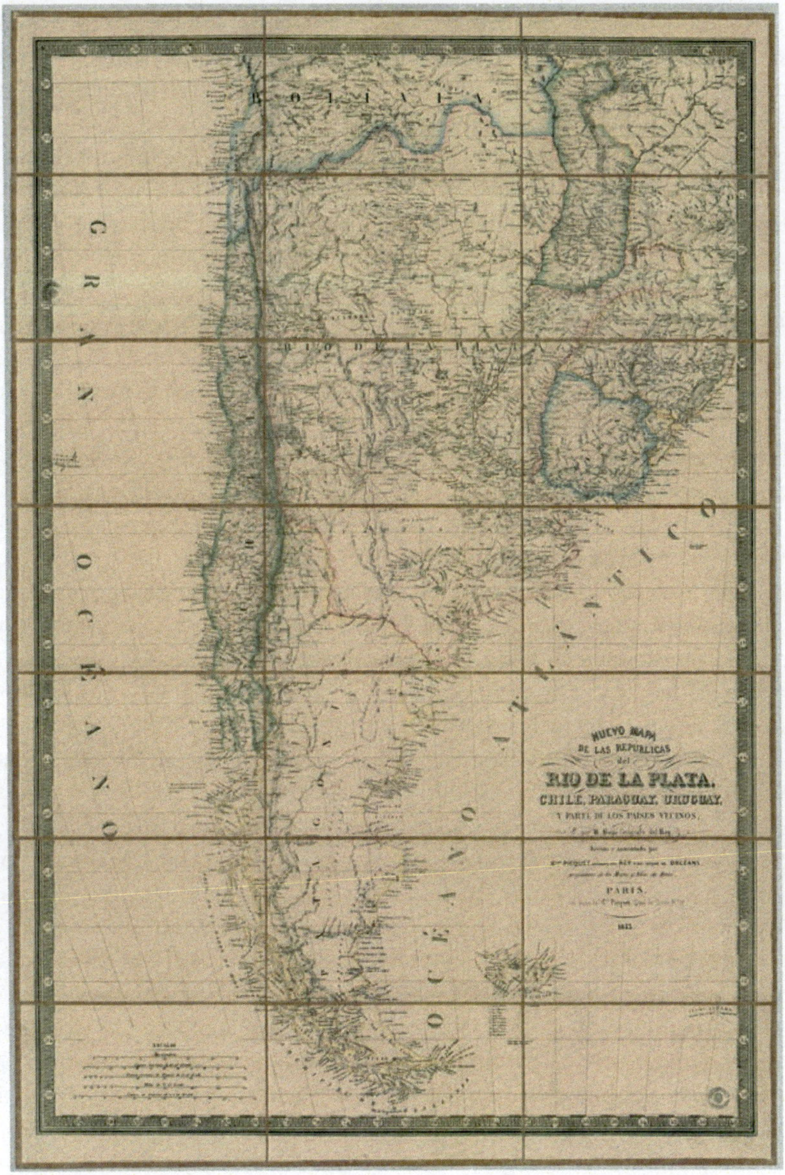

Fig. 5.5 Brue *Amerique Meridionale* (1843)

power".[20] The atlases published by the Colton family between 1855 and 1879 represent the jurisdiction of Chile to the west of the Andes and up to the Magellan Strait (western Patagonia) and that of the Republic of Argentina up to the 40° south latitude, although they acknowledge the territorial claim of the latter over Patagonia. There is no international boundary on the Andes and the Malvinas Islands do not appear on these maps.[21]

The production of English cartographers and editors—the Arrowsmiths—deserves special mention, particularly the map of southern America, which inspired cartographical representation during the central decades of the 19th century. Produced on the information gathered between 1824 and 1832 by the first British Consul in the United Provinces of the Río de la Plata, Woodbine Parish (Parish 1958),[22] and on the reports and maps of Parker King and Fitz Roy's surveys (1826–1836), it shares some characteristics with maps of that period. It clearly shows great care in improving the outlining of the coastal profiles and communication routes and in the localization of commercial interest resources according to the reports provided by scientific expeditions (Llanos Reyes 2010). The map of South America published in "*Voyages of the Adventure and Beagle*" (1839) and in the 1842 edition of *London Atlas of Universal Geography* considered the only Hispanic-"criollo" settlement that survived at the mouth of the former and the reconnaissance carried out by B. Villarino (1782) and during the military campaign of J. M. de Rosas (1832–1835). Thus, it represents Chilean jurisdiction up to Chiloé and, to the east of the Andes, the United Provinces up to the Negro and Limay rivers. The Patagonia—referred to as "Inhabited by Wandering Tribes of Indians"—stretches to the south of the Negro River. The map shows also in red Guayana, La Plata and Falkland, potential areas of interest or influence for the British[23] and, in detail, with a bigger scale in three marginal frames, Patagonia and Tierra del Fuego, Falkland and Chiloé. The number of toponyms on the coasts contrasts with the blanks in the interior of Patagonia and Tierra del Fuego, except for the basin of the Santa Cruz River, explored by Fitz Roy in 1834. In the Falkland and Georgia archipelagos, toponymy and legends indicate English discovery and exploration antecedents—*Falkland Is. Discovery by Sir Hawkins 1574 and Land discovery by La Roche 1675/I. Georgia of Cook 1775*. The map constructed by J. Arrowsmith and drawn by A. Peterman (1852) to be published in the first Spanish edition of W. Parish's work reinserted the southern territories within lateral frames, depicting just the outline and including the names of some ports and geographical features, and in the interior, devoid of indigenous references but for a couple of rivers in the Atlantic slope (Lois 2007: 109) (Fig. 5.6).[24]

Apart from the political delimitations that suggest southern territories representations as separate entities or integrated—either fully or partially—to the States in formation, cartography enables other ways of conceptualizing space. Whereas some maps offer the image of an indigenous territory, occupied by a diversity of politically and territorially sovereign nations (SDUK 1838; Radefeld 1844; Rapkin 1851; Colton 1856; Garnier 1860), others show an empty/ied Patagonia, with few indigenous names and legends, which conveys the idea of a barren and deserted space, a *terra nullius* (Lizars 1831; Tanner 1836; Malte Brun 1837; Bradford 1838; Greenleaf 1840; Arrowsmith 1842; Johnston 1861) (Fig. 5.7).

Post-revolutionary governments, state-building agents and new territorial actors collected information to justify their authority and support their territorial claims on the grounds of the 1810 *uti possidetis* principle. Likewise, they sought to legitimize their claims, projects and interventions on the basis of cartography technical-scientific knowledge.

Fig. 5.6 Tanner *Chile, La Plata and Montevideo* (1836)

French naturalist Claudio Gay was hired to explore, systematize geographical information and draw up an atlas with maps of all Chilean provinces. The *Mapa para la inteligencia de la "Historia física y política de Chile*", published in Paris in 1854,[25] was based on Spanish and English cartography, especially elaborated by Malaspina, Parker King and Fitz Roy's expeditions, updating data and broadening the knowledge of the interior as a result of the voyages from Atacama to Chiloé, between 1830 and 1842 (Sagredo Baeza 2018). By representing "Chile itself", the "country inhabited by the Araucanians" and, in a marginal frame, the outline of the southern territories to the south of the 42°, Gay initiates the process of building a territorial image that corresponds to boundaries established in section one of the 1833 Constitution: "The territory of Chile extends from the desert of Atacama to

Fig. 5.7 Peterman *The provinces of Río de la Plata and adjacent countries (1852)*

the Cape Horn, and from the Andes mountains to the Pacific ocean, including the Archipelago of Chiloé, all the adjacent islands, and those of Juan Fernández".

Fig. 5.8 Gay Mapa de Chile (1854)

The Argentine Confederation, after the secession of Buenos Aires, also started cartographical and topographical registers by the mid-1850s by hiring a French physician—J. A. V. Martin de Moussy—who had submitted a vast exploration plan to elaborate a geographical, ethnographical and historical work to President J. J. de Urquiza (González Bollo 1999). The Atlas published in Paris in 1865 and funded by the president of the united republic—Bartolomé Mitre—inspired all subsequent official maps, which attempted to counteract the foreign viewpoint. Although it amounted to the first national cartography, it rather constitutes "a graphic set that provided a fragmented whole of Argentina in the 19th century" (González Bollo 1999). Martin de Moussy represents a territory of effective state sovereignty comprising 14 provinces, formally incorporating the indigenous territories of Gran Chaco and La Pampa and describing Patagonia as "an almost unknown separate territory", claimed as part of the colonial heritage.[26] Following the configuration of Parish-Arrowsmith-Peterman's maps, he defines the "Southern Indian Territory" from the actual frontier to the south of Buenos Aires, Mendoza, San Luis and Córdoba to the Negro and Limay Rivers.[27] The chart that represents the territories located to the south of said rivers offers a territorial history narrative that highlights the "ancient Missions" of the Lake Nahuel-Huapi (41° south latitude) and the "fabulous City of the Caesars" (43° south latitude) as early antecedents of the Spanish presence. Also, it represents the colonial foundations on the Atlantic coasts—"La Piedra colony (1779)" on San José Peninsula (currently Valdés Peninsula), "Viedma colony (1780)" at Deseado Port and "Viedma colony (1781)" at San Julián Port—indicate the effective occupation of the region and the beginning of colonization. Additionally, the settlement Carmen de Patagones and the Chilean colony of Punta Arenas are depicted as the unique inhabited site by the mid-19th century.[28] The *Description géographique et statistique de la Confédération Argentine* (1860–1864) includes remarks that refute the negative image of Patagonia, stating that according to what natives affirm, the conditions of the interior are more favorable than those of the maritime coasts known by Europeans. It evidences the unawareness of the interior of Tierra del Fuego, confirming it through a cartographic inscription that describes the Isla Grande as a mountainous territory and only inhabited on the coasts. It observes that the possession of the Patagonian region belongs to the heirs of the Crown of Castile since Spanish subjects were the first explorers. However, Martin de Moussy clarifies that "the boundaries of the Confederation have not yet been settled clearly by means of treaties with the neighboring States". Whereas the Atlas identifies the Malvinas/Falkland archipelago with double toponymy—French and English—and fails to identify any settlement, the *Description* refers to the discovery and exploration antecedents of the Dutch, English and Malouines, mentions the occupation of France and Spain, the jurisdictional acts of Buenos Aires government in 1820 and the English occupation in 1833: "*Le gouvernement argentin, successeur de l'Espagne et héritier de ses droits, y envoya, en 1820, un gouverneur, M. Vernet. Celui-ci, en 1833, fu dépossédé purement e simplement par l'Angleterre, qui prétendit que l'Espagne les lui avait cédées en 1771*".[29] Silence over the territorial dispute would evince a policy that, according to Ferns (1984), was due to the decision of not rendering the Malvinas issue into a serious conflict.

De Moussy Carte de la Patagonie et des archipels de la Terra de Feu des Malouines (1865)
http://bdh-rd.bne.es/viewer.vm?id=0000000379

5.3 Competing for Resources and International Rivalry

The failure to find new template lands in the South Pacific Ocean and the confirmation of the uninhabitable conditions of those lands beyond the 60° south latitude put an end to the expectations to find the alleged southern continent. Following J. Cook's voyages during the 1770s, the exploratory interest shifted to the assessment of geographical coordinates and deepening the knowledge of the ports that served as ports of call for navigation and bases for the hunting of sea mammals.

France and England, that had taken possession of the Malvinas Islands and founded settlements in 1764 and 1766, respectively, temporarily abandoned the occupation plans of southern lands. After acknowledging the Spanish sovereignty on the archipelago pursuant to the Family Compact between the Bourbons, in 1767 France yielded the precarious premises at Port Louis (East Malvina) and at that port, renamed *Puerto Soledad.* The Spanish colony and prison survived until 1811 (Levaggi 1978). In turn, the English who had been expelled from Port Egmont (West Malvina/Saunders Island) in June 1770 by an expedition, sent by the governor of Buenos Aires, recovered the settlement after diplomatic negotiations. However, they dismantled it—without waiving their rights and leaving occupation signs (Rose 1929)—due to the fact that its maintenance was too costly and its site appeared to be inconvenient.

Nevertheless, increasing competition for controlling maritime routes and exploitation of marine-faunal resources urged the exploration of the American insular tip and new regions to the South Pole. Within the context of growing demand for fat and oil, due to the process of industrialization started in England, Cook's reports on the large number of pinnipeds and cetaceans promoted the economic valorization of the southern region. The coasts of Patagonia, Tierra del Fuego and the sub-Atlantic islands became key areas for sealing. Thus, as the seal population declined due to intensive seal hunting, competition forced hunters to scout for new hunting sites, broadening the space of interest to the sub-Antarctic and the known coasts of the Antarctic Peninsula (Zarankin and Senatore 2005).

Until the end of the 18th century, the technological limitations of navigation made it risky to sail beyond the polar circle. Conversely, technological advances and guaranteed economic benefits through oil and fur trading since the early 19th century triggered the rapid growth of hunting. Unlike previous periods, both the States and private enterprises operating simultaneously undertook the exploratory labor on the islands in the Indian Ocean, the coasts of Patagonia and the South Atlantic archipelagos (Senatore and Zarankin 1999).

Maps reflect this knowledge development by identifying new exploitation areas. Although there were no state jurisdictions, toponyms reveal the leading countries in a competition in which the English, French, Russians, Dutch, North Americans and companies based in the Río de la Plata took part.

The first maps of the South Shetland Islands resulted from the intermingled interests of sealers and officers of the empire. In 1819, Captain E. Bransfield of the Royal Navy surveyed and mapped the islands in an expedition organized from Valparaiso British station. Captain W. Smith, trader and sealer who regularly traveled between this port in the Pacific and the port in Montevideo, had spread the news of the existence of new islands near the Antarctic Circle. E. Bransfield's map, as well as that of G. Powell, both published in 1822,[30] shows the interest in gathering information in order to consider the possibility of settling a colony. As Basberg states (2017), since the abandonment of the Malvinas/Falkland Islands, British sealers—facing North American competition—had no operational base near the hunting sites.

J. Weddell's voyage account, dedicated to the First Lord of the Admiralty and published in London in 1825, attempts to show the significance of the southern discoveries for Great Britain by describing the voyage carried out between 1822 and 1824 and stating some observations of his two previous voyages, between 1819 and 1821. It is an important source in order to learn about the expedition routes, the hunting sites and intensity, in a highly competitive time due to its high profitability. From the Malvinas Islands, he headed to the South Orkney Islands, and sailed to the South Sandwich Islands through a "free-of-ice sea". In February 1823, he reached the 74° 15′ south latitude, some degrees beyond that of James Cook (71° 10′ south latitude). On his way back, after sheltering at South Georgia, he wintered at the Malvinas Islands and provided a detailed description of the place. During the last stage of his journey, he sailed round the South Shetland Islands and the area around Cape Horn, an area of which—according to Weddell (2006)—there were no accurate descriptions. Three maps show the itineraries at different scales—hemispheric, regional and local—enabling a glimpse of the economic, political and scientific interests that fueled the progress over the Antarctic continent.

The hemispheric map reveals the global imperial project by highlighting the positioning of Great Britain in the southern hemisphere and proposing a circumpolar territoriality. On the one hand, it declares the British leading position in the discovery process, by stating that Cook reached 72° south latitude and the conquest of the most southerly latitude of 74° with the legend "Clear Sea in 1823 (Weddell)". On the other hand, the spatial framing at 40° south latitude allows the visualization of a possible access to the South Pole from Tasmania and New Zealand, in the Indian Ocean, and the South Atlantic Ocean from the insular tip of South America.[31]

The chart that shows the routes of the three voyages of Weddell provides an image of the south Atlantic region that stretches from Patagonia to the Antarctic lands and adjacent seas. Far from the alleged scientific neutrality, expressed in the title "Voyages of Investigation", it territorializes sites of interest by using main toponyms (uppercase letters) and including important information. The legend "Land not accurately known" appears next to the toponym—TRINITY LAND—placed on the recently discovered coasts of the Antarctic Peninsula. The biggest toponym, "THE SEA OF GEORGE THE FOURTH", appearing at the most southerly latitude ever reached (74°), is of political nature. Both the qualifying term "Navigable" and the reference to the sighting of a huge number of whales and fowl, note the chances of successfully operating in the area.[32] Lastly, the chart of South Shetland, main target of his third voyage, evinces the exploratory advance through using many toponyms and indicates that William Smith discovered these islands in 1818 by including a legend.[33]

The quest for convenient ports for docking and provisioning of the ships that sailed around Cape Horn was a constant target of English expeditions. After exploring the southern coasts of Tierra del Fuego and Staaten Land, by the end of 1774 and the beginning of 1775, J. Cook suggested Port Christmas, in the southern tip of the Isla Grande de Tierra del Fuego, and Port Years, on the northern coast of Staaten Land.[34] Captain James Colnett, who had shared Cook's voyage, also suggested Staaten Island for settling a base for hunting whales after the expedition he led, serving in the Admiralty between 1792 and 1794.[35]

The hydrographic labor developed by Antonio de Córdoba (1784–1785) aimed at comparing the feasibility of sailing through the Magellan Strait to sailing round Cape Horn. He ratified the Nodal brothers' recommendation of abandoning the risky navigation through the strait and sail along the route to the south. This appreciation coincided with the importance assigned to the ports mentioned by Cook due to the availability of wood for fuel, fresh water and fresh food. Córdoba manifested his preference for Puerto de Año Nuevo, whose lands were adequate for cattle breeding which could be transported from the Malvinas Islands and, unlike Good Success Bay, avoided sailing through the Strait of Le Maire.[36] Cartography identifies the best Atlantic ports along the route round Cape Horn by presenting a detailed depiction in marginal frames and bigger scale. Both Cook's chart of the southern extremity of America (1777)[37] and the map that was published together with Córdoba's voyage account (1788) present the plan of Puerto de Año Nuevo, although the latter also includes those of Soledad and Good Success Bay,[38] idea that reappears in W. Faden's work (1802) (Fig. 5.8).[39]

The exchange between Viceroys Croix and Loreto during 1788 reveals the concern for confirming the news of a foreign occupation on Staten Island and the existence of an English colony in the so-called New Ireland, located at 55° 45′ south latitude, according to the statements of an English Captain to the Governor of Malvinas Islands, Ramón de Clairac.[40]

During the 1820s, news of the existence of an English settlement on Staaten Island reappeared. According to descriptions included in atlases and universal geographies, Hopparo colony had been founded by English sealers in 1818. It had a fort and hundreds of people (400 or 500) among soldiers and colonists.[41] As it was probably one of the many temporary settlements of sealers and whalers, official cartography failed to represent it and Great Britain never claimed it. Instead, it appears on German and French non-scientific maps: Reichard (1804),[42] Buchón and Carey (1825), Weinland (1828) and Schlieben (1830),[43] among others.

C. Weiland's map (1828) represents the sealers' area of action, an international maritime territory identified by the two main toponyms *Stilles Meer* and *Süd Atlantischer Océan.* It is one of the first maps to use the name "South Atlantic" Ocean to refer to an economic space that stretches from the Río de la Plata—strategic site for hunting activity—up to the 60° south latitude. This map included the various capture areas: the coasts of Patagonia and Tierra del Fuego, the Malvinas/Falkland Islands, Shag Rocks, South Georgia and South Sandwich, and "the South Pole lands": Orkney, Shetland and Trinity Land. By describing the latter as "under the spell of a frozen area", it expresses the harshness of a freezing weather in times when the profits of hunting failed to compensate for the harsh working conditions. After several decades of intensive exploitation and with a peak of highest earnings in the 1820 s, the sealers cycle was already declining (Fig. 5.9).[44]

Fig. 5.9 Weddell *The Polar part of the Southern Hemisphere* (1825)

Hydrographic surveys developed by the Admiralty between 1826 and 1836 served the purpose of broadening the knowledge of the coast of Patagonia and the insular tip.[45] Having been partially surveyed in previous Enlightenment expeditions performed in the 18th century, Weddell—in his accounts—calls our attention by describing its ports as ideal as ports of call and with availability of wood and fresh water. Fulfilling these aims required the systematization of the knowledge produced by previous expeditions[46] and particularly, the practical knowledge of pilots with local experience (Martinic 1999; García 2010). The accounts of P. Parker King (1826–1830) and R. Fitz Roy (1831–1836) described contacts with English and North-American sailors and sealers who provided information and in some cases served as guides when sailing through the Magellan-Fuegian channels and in the reconnaissance of the coast of Malvinas/Falkland Islands (Parker King 2015; Fitz Roy 2013). As Passetti remarked (2014), these experts while sailing to the south were also transmitters of the spatial knowledge provided by native people, with whom they regularly exchanged fresh victuals (guanaco meat and fish), furs and information in return for metal, garments and alcohol. The same author highlights the mutual dependence between sailors-hunters and English sailors-explorers. The former needed the maps made by the latter and the latter, in turn, could keep their role and prestige within the Royal Navy, thanks to the pressure exerted by trading and hunting companies for the official funding of scientific expeditions. Cooperation and trust relationships between both is clear in mail, maps and accounts delivered for the Admiralty before the ships returned to England (Passetti 2014: 265-266). The reports compiled by Fitz Roy and published in 1839 account for the effectiveness of the combined effort, and offer plenty of information on sailing conditions, routes and ports, economic expansion areas and territorial occupation alternatives.[47]

Cartography produced during the surveys was published upon the completion of the various stages. The maps of the southern tip of America[48] and of the Magellan Strait[49] were published after Parker King's hydrographic labor and the charts of the Malvinas/Falkland Islands[50] and that of the southeast of Tierra del Fuego[51] were published after the second campaign. The most significant information provided by the *Chart of a part of South America*, published initially in 1831 by John Murray for the *Journal of the Royal Geographical Society* is the advance in the knowledge of the geographic configuration of the archipelago of Tierra del Fuego. On the one hand, it sheds light on the geographical enigma that had lasted a century by determining the inexistence of San Sebastian Channel and, on the other hand, it introduces new maritime passes, Beagle and Murray Channels. Toponyms have a key role on the chart. They show a juxtaposition of Spanish, Dutch and French names and the prevailing English names. The most important ones (bigger) correspond to ports and interesting features for navigation. By keeping the name suggested by Narborough for the Isla Grande de Tierra del Fuego, "King Charles South Land", it expresses the continuation of the British interests in the southern region. Although the instructions from the Admiralty were to privilege native names, appreciating their descriptive nature, these were replaced by others with no physical or historical connection to the places. As Passetti (2014) stated, Fitz Roy was particularly keen on changing

Fig. 5.10 Selma *Carta esférica de la parte sur de la América Meridional* (1788)

the names of the places he visited and immortalizing the names of his relatives (Fig. 5.10).

During the following decades, the Admiralty Hydrographic Office re-issued the maps. The reproductions produced since 1838 by the *Depósito General de la Marina* (Paris) and since 1860 by the *Dirección de Hidrografía* (Madrid), as well as private editors working in Europe and the United States helped to spread those maps (Martinic 1999). Considered as scientific cartography, these maps were the models on which the charts of Patagonia and the Magellan Strait were drawn, and through the strategy of re-naming they started a process in which all references of indigenous spatiality faded into oblivion.

While broadening the hydrographical and geographical knowledge, international competition over the use of interoceanic routes and the exploitation of maritime resources increased. New independent governments also appeared on stage in order to confirm their territorial claims. In 1820, the Supreme Director of the United Provinces ordered the occupation of the Malvinas Islands but, due to the difficulty of affording a permanent settlement, the authorities in Buenos Aires just granted some

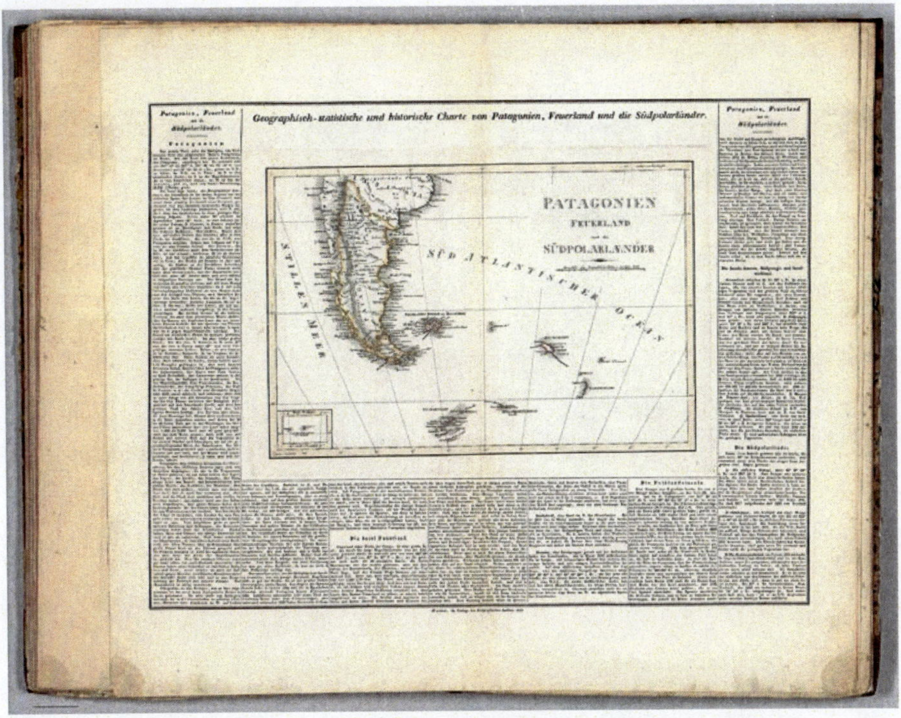

Fig. 5.11 Weiland *Charte von Patagonien, Feuerland und die Südpolarländer* (1828)

lands and permits to perform economic exploitations (hunting of marine mammals and cattle breeding). Only in 1829 Luis Vernet—one of the first concessionaires of Soledad Island—was appointed Commander of Malvinas islands by the government in Buenos Aires and was commissioned to establish a colony within three years. In return, he was granted the exclusive fishing rights for 20 years in the Malvinas and the coasts of Patagonia to the south of the Negro River. Vernet attempted to involve Consul W. Parish in the enterprise in order to promote the arrival of settlers (Caillet-Bois 1982). He drew a map of East Malvina on the basis of the reconnaissance performed between 1826 and 1828. He propounded a division of the island into 11 sections with lands suitable for the development of cattle breeding and an area reserved for the state.[52]

The creation of Malvinas Political and Military Command on Malvinas[53] and the subsequent jurisdictional acts of Buenos Aires government triggered the expansive plans of the United States and England over the South Atlantic Ocean. By the end of 1831, Vernet seized three American ships. As a result, the sloop "Lexington" raided Puerto Soledad as a reprisal for the embargo on a consignment of sea lion skin. The commander traveled to Buenos Aires to defend his case and elaborated a report that offered the first Report on "the issue of the rights of Argentina over the Malvinas Islands and adjacencies and over the coasts of the continent up to Cape Horn".[54] On due reflection on the European means to legitimate their territorial appropriations,

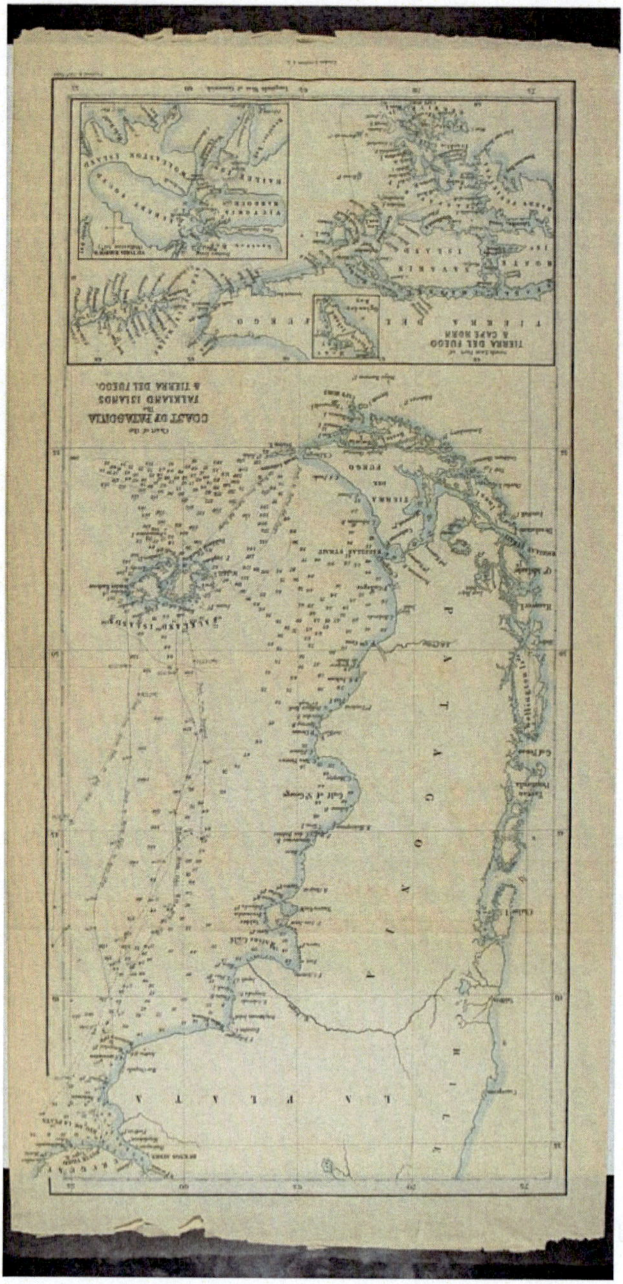

Fig. 5.12 Parker King *Chart of a part of South America* (1831)

Fig. 5.13 Vernet "*Plano geográfico de la isla oriental de Malvinas*" (1829)

he asserts that, apart from the discovery and the effective and permanent occupation, "a nation may acquire dominion over a territory that had belonged to another nation, either by inheritance, cession, sale, exchange or treaty". After presenting the antecedents that supported the physical possession of the Malvinas Islands by Spain, he argued that the independent government succeeded to all rights on the principle that "all lands that would have belonged to the Viceroy or Captaincy which it had succeeded (except for later agreements) correspond to each state, big or small, and also all exclusive rights pertaining to such territory". The occupation of the islands by the Republican government was made legitimate by adopting this ruling, so Vernet argued that the jurisdiction defense was due under the law of nations.

While Buenos Aires and Washington were immersed in a dispute that ended in the breakdown of diplomatic relations and Argentina was trying to attract settlers to the colony founded by Vernet, the Admiralty sent the sloop "Clio" as an "act of sovereignty" on the Malvinas Islands. Captain J. J. Onslow ordered the eviction of the Argentine garrison on 2 January 1833, and the following day, he took possession of the islands.

Although from the end of 1820s a group of parliamentary members attempted to reduce the naval costs, influential agents—diplomats, officials of the Royal Navy and traders—promoted the reoccupation of the Malvinas (Warnick 2008). British Consul in the Río de la Plata, W. Parish, based on the information provided by Vernet, wrote an article describing the favorable conditions of the islands for colonization and crew provisioning, highlighting the abundance of fish, wild cattle and anti-scurvy food (Parish 1833). The attached map, based on Vernet's "Geographical Chart of East Malvinas", shows an empty space with few names, an image that shows the lack of a concrete project of colonization and proves to be functional for the use of the islands as a naval base and convenient port for commercial activities in South America (Fig. 5.11).

Buenos Aires authorities raised claims through its Minister in London[55] and, although they were violent and threatening at first, the disputed gradually turned to be an accepted element within Anglo-Argentine relations (Ferns 1984). According to Warnick (2008), the financial dependency and the incapacity of paying the loan to the Baring Brother fostered the British indifference toward Argentine protests, which escalated at the beginning of the 1840 s upon formal colonization and the establishment of a civil government in the islands.

Great Britain would have acted in reaction to Buenos Aires claims and the increasing French and American competition rather than defending its old rights, since the occupation of the islands was performed with no coordination of the different governmental areas responsible for the decision-making in colonial and foreign policy. They also failed to assess the conditions for the administration of such a far and peripheral possession (Warnick 2008). However, the possibility of having a base in the South Atlantic Ocean appeared to be a logical option of the imperial strategy that, since the end of the 18th century, combined the efforts of private enterprises and the Admiralty services in providing infrastructure and operative support in order to reach the economic targets (Jones 1981).

Fitz Roy, who recognized the Malvinas/Falkland Islands as the most logical site for establishing a port for repairs and provisioning, justified the imperial project with an inconsistency in Buenos Aires claims on the grounds of territorial inheritance and affirming British rights on the archipelago. "Do discovery, first occupation and settling on new and uninhabited lands grant right to possession? If that is the case, Great Britain is the legal owner of the islands. Davis was the first one to discover them; Strong was the first one to disembark on the islands and, except for the French, Byron was the first one to take formal possession of them. Again except for the French, Macbride was the first one to colonize them". Later on he keeps arguing: "England never acknowledged Spain's claim as valid. And France having left the islands, the issue was now between Spain and Great Britain. Spanish did not discover, disembark and settle on Falkland before the British and Spain's only claim rests on the unstable grounds of a Papal Bull by which Spain could claim Tahiti, the Sandwich Islands or New Zealand" (Fitz Roy 2013: 213). In the "Narrative", he also manifested his concern over the lack of immediate initiatives for the exploitation of resources and the lack of a regular authority on the islands (Fitz Roy 2013: 216, 246).

In fact, the British presence was quite precarious during the first decade of occupation. There were no state or private investments to support the colonization started by Vernet and they just kept a small naval garrison that had minimum support from the central government. Concurrently, cartography enabled an intellectual, discursive and visual appropriation of space by representing the South Atlantic from the 50° south latitude up to the Antarctic Polar Circle as English territory. The map *The Falkland Island* (1834), produced and published by J. Arrowsmith, reveals the results of Fitz Roy's surveys and, although some coasts were drawn with dotted lines showing the lack of enough information, it was the most accurate map of the archipelago for several decades. Toponymy and legends indicate English rights, and the map of the world included in the bottom right corner represents the Malvinas with the same color as other British islands.[56]

Maps published by the Society for the Diffusion of Useful Knowledge (SDUK), an entity founded in 1826 to produce and distribute books and maps of high quality and accessible to a public interested in improving informal education (Cain 1994) served an ideological purpose by shaping a territorial representation of the empire. The map "*Falkland Isles o les Maloines*" (1838), published some years after the occupation, promotes the potential of the islands with the following legend: "These islands have been colonized and abandoned by France, England and Spain consecutively but are now part of England possessions. They have great ports, good soil, plenty of cattle, rabbits, hunting and fishing; and although there is no wood, there is a lot of peat for fuel. The highest and lowest temperatures recorded are 76° to 25°" (Fig. 5.12).[57]

Since the 1820 s, the French Navy also considered projects to recolonize the Malvinas Islands or having a settlement in southern Patagonia. Commander Dumont d'Urville, who used Puerto del Hambre as a base of his Antarctic operations in 1837, noted Sarmiento de Gamboa's right choice for founding a colony. Another member of his expedition, C. Vincendon Dumoulin, proposed the occupation of the Magellan Strait within the plan of strengthening France's position in Oceania, following the incorporation of New Zealand by Great Britain (Berguño 2002). But that same year the government of Chile took possession of the strait, and French authorities, more interested in opening the Panama Isthmus for interoceanic communication than in colonizing Patagonia, acknowledged Chilean rights and gained the possibility of operating in the Pacific Ocean from the Valparaiso port.

Exploration progress and cartographical production show the "inter-play of political, economic and cultural practices that converged to define a maritime space as an object of scientific inquiry" (García 2010: 285). They offer a glimpse of strategic and commercial interests that explain the inclusion of the South Atlantic Ocean in European geopolitical agendas.

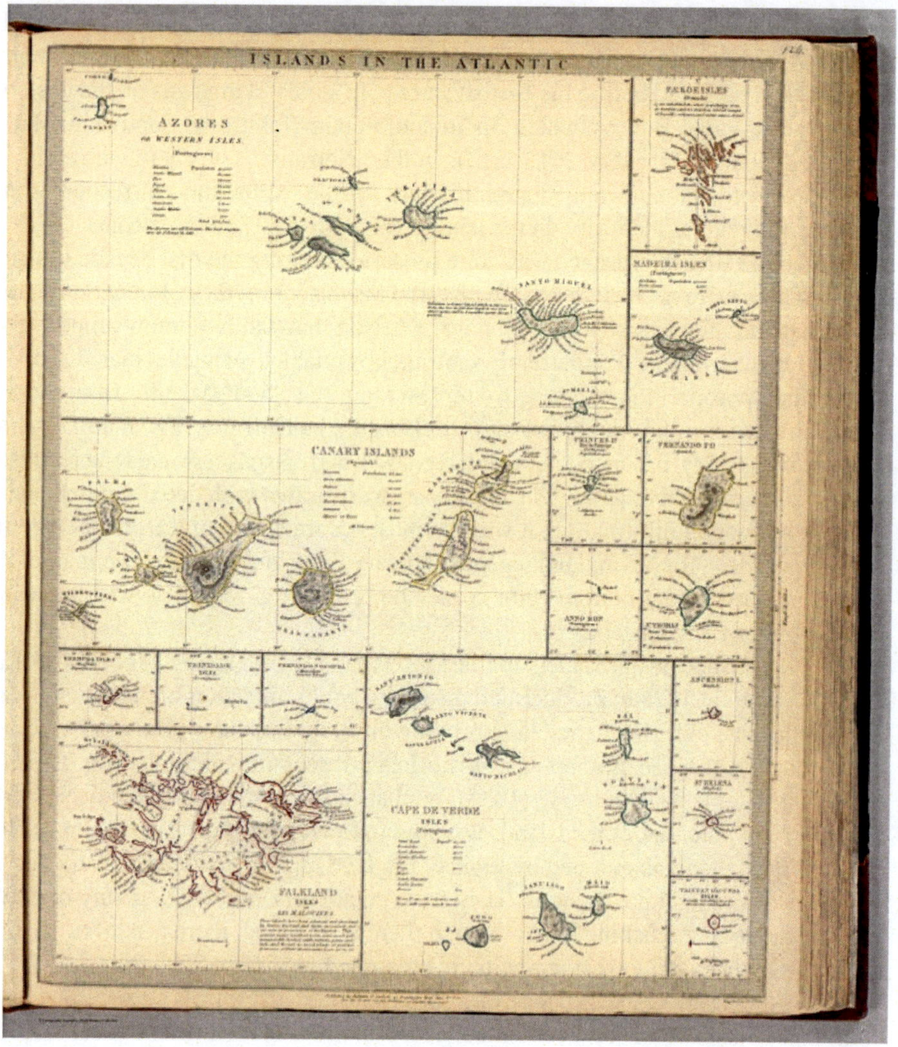

Fig. 5.14 SDUK *Falkland Isles or Iles Malouines* (1838)

5.4 The Circumpolarity of the Southern Pole: Imperial Propaganda?

Changes in cartographical representation of the southern hemisphere in the first half of the 19th century accounts for the construction of new spatial imaginaries during the early Antarctic exploration,[58] within the context of strong economic competition and expansion of British interests on a global scale.

The representation of Antarctica—undoubtedly the most important transformation—amounts to the last stage in a debate that had lasted a century, over the existence, configuration and territorial nature of the unknown southern lands. Like previous stages, maps show the manners in which they interpreted new evidence and, as Lois stated (2018), they show the observer—his speculations, hypothesis and beliefs over what is possible or impossible—rather than the geography postulated as referent. Thus, while the main English cartographers opted for considering the evidence as insufficient to represent the alleged continent, a small group of German cartographers drew the whole outline by joining the known coasts. Other American cartographers, for their part, assumed an intermediate posture and accepted the continent but outlined just the known areas and included the toponym "Antarctic Continent". Ch. Wilkes was the one who suggested this name, during the United States Exploring Expedition (1838–1842), taking into account the vast mountainous land found lying between the 64° and 67° south latitude and the longitudes 154° and 79° (Hobbs 1940; Bulkeley 2016).

Apart from the proposals of the possible configuration of the fifth part, another new element in cartographic representations of the southern hemisphere is the concept of *circumpolarity*. Used since the 17th century to control the lands surrounding the Arctic,[59] it appears in the south of 19th century maps, possibly as a result of the interpretation of the data provided by J. Cook's circumnavigations, particularly the second one, mirroring knowledge production ways used in the configuration of the northern hemisphere. Weddell's map "The Polar Part of the Southern Hemisphere" (1827) presents a frame of the southern hemisphere at the 40° south latitude and, as the title clearly states—"Exhibiting a comparative view of Land & Sea with that of the Northern"—presents a "view" of a southern circumpolarity mirroring that of the northern hemisphere.

Being increasingly more explicit in cartographical titles and legends, the concept led to the distinction of the regions to the south of the Antarctic Circle, whose identity started to be unveiled from the new science protocols. Engraver S. J. Neele and cartographer J. Wilkes' polar projection map "Countries surrounding the South Pole" (1807)[60] represents the southern hemisphere up to the 38°, depicting the southernmost discovered lands describing Cook's itineraries and those of other previous voyages. It also gives place to assuming the existence of a landmass of significant size by representing the "icefields" between the 60° and 72° south latitude.

N. Desmarest published in France an Atlas of Physical Geography that includes a "*Carte des glases, circompolaires australes*" (1827),[61] whereas in England, the Society for the Diffusion of Useful Knowledge (SDUK) and Arrowsmith's company published polar projection maps with a similar conception, such as "Circumjacent the South Pole" (1831)[62] and the map included in the map of the world "The World on Mercator projection" (1832) that depicts the lands that are contiguous to the South Pole—"Lands contiguous to the South Pole".[63]

Bearing in mind the resources used by English cartography to set the spaces under their control and those of imperial interest, it is very likely that the concept of circumpolarity in the southern hemisphere, mainly oceanic, was part of appropriation, legitimation and territorial propaganda strategies. Also, it would enable them to express their territorial aspirations over a new southern frontier and show control over the access to the alleged Antarctic continent from the British possessions in

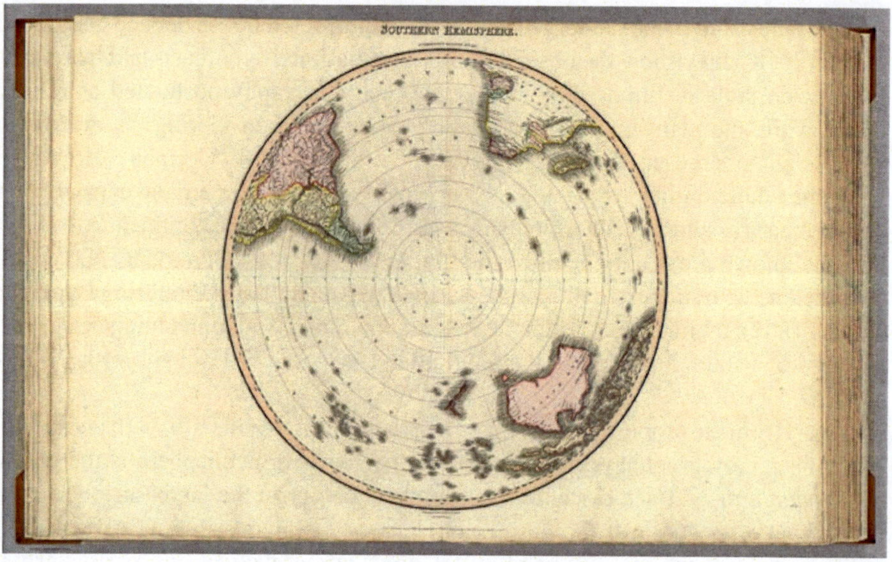

Fig. 5.15 Pinkerton *Southern Hemisphere* (1812)

Australia and New Zealand and, in the borders of the Atlantic Ocean, from the islands in the southwestern Atlantic Ocean and Cape Colony.

Polar projection maps included in the general atlases of Edinburgh School, J. Pinkerton (1812)[64] and J. Thomson (1814)[65] represent the southern hemisphere from the Equator to the Pole, depicting just the known lands and locating a frozen sea—Ice Sea Antarctic—to the south of the Antarctic Circle. While Thomson manifests "scientific neutrality" by using color in a homogeneous manner for each continent and omitting political divisions, Pinkerton's map differentiates domains and territories under British influence by means of colors (Fig. 5.13).

Nationalistic and didactic maps published by the SDUK, Wyld's cartographical company production and the well-known map of the world of H. Smith Evans provide visual evidence of the imperial power. They also highlight with the same red color Great Britain and its possessions, and overseas territories considered of interest. In the South Atlantic Ocean, the Falkland Islands, the sub-Antarctic archipelagos—*Isle of Georgia, South Shetlands, South Orkneys*—and the known coasts of the Antarctic Peninsula—*Graham Land* and *Palmer Land*—constitute one of the circumpolar regions over which British territoriality in the southern hemisphere is conceived (Fig. 5.14).[66]

Wyld's maps *Southern hemisphere* (1849)[67] and *Chart of the south circumpolar regions* (1844) published by J. D. Hooker, a botanist in J. Ross' expedition,[68] correspond to the peak of British interests' expansion and show the coincidence in the advance of geographical and imperial frontiers in the southern hemisphere. The latter clearly exposes the leading position of the British exploration of the regions closest to the South Pole, representing the discoveries of Cook (1773–1775), Weddell (1822–1824) and Ross (1839–1843) (Fig. 5.15).

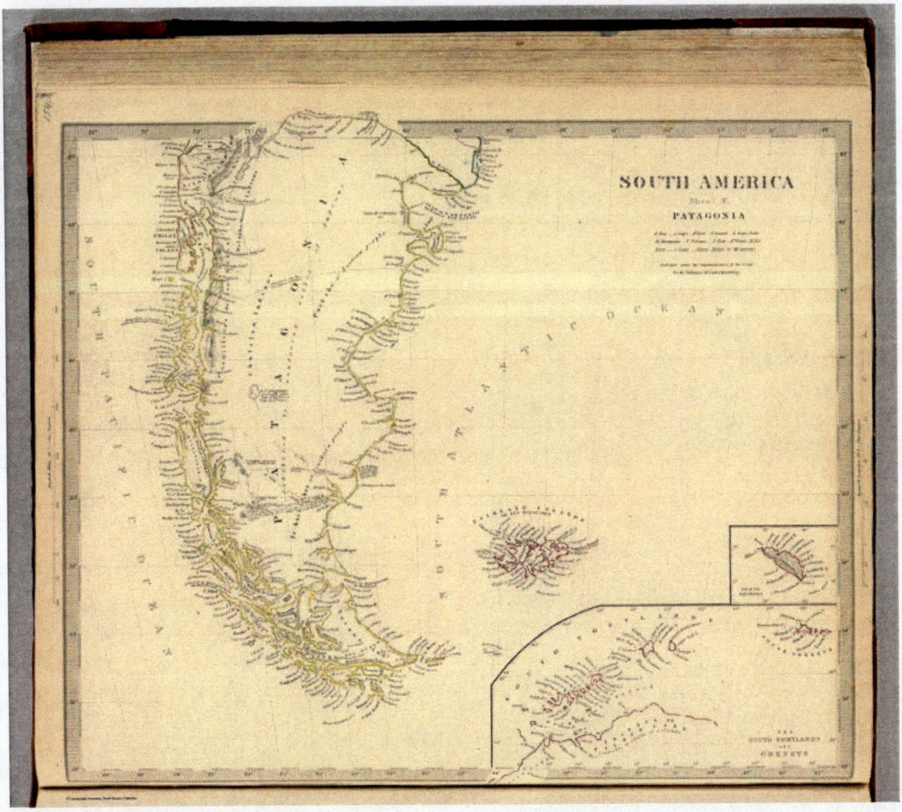

Fig. 5.16 SDUK *Patagonia, S. Shetlands, S. Orkneys* (1838)

Evans Smith's Mercator projection map of the world "Emigrations map of the Word" (1852)[69] depicts northern and southern hemispheres with the same extension in latitude and highlights British possessions in red. To the south of 60° south latitude, the few known coasts fuels the desire for lands yet to discover, and the empty spaces, "closely related to imperial fantasies" (Yan 2007), are used to include tables, references and a symbolic element that summarizes the underlying ideology: the image of a ship sailing across the globe and the legend "Britain upon whose empire the sun never sets". Another feature of this colonial map is that its center is the Pacific Ocean and it represents Europe and Africa twice, which allows a global and uninterrupted vision of maritime circuits that connect British territorial annexations during the 1830s and 1840s.

The cartographical construction of the southern circumpolarity, linked to the necessities of a colonialism that aims at controlling strategic sites on the planet, would constitute a way of symbolic appropriation of territories in advancing to the Southern Pole. Likewise, it would be an expression—like others—of the articulation of science, politics and commerce in the creation of categories that constitute epistemological frames and, concurrently, it nourished the geopolitical imaginaries of

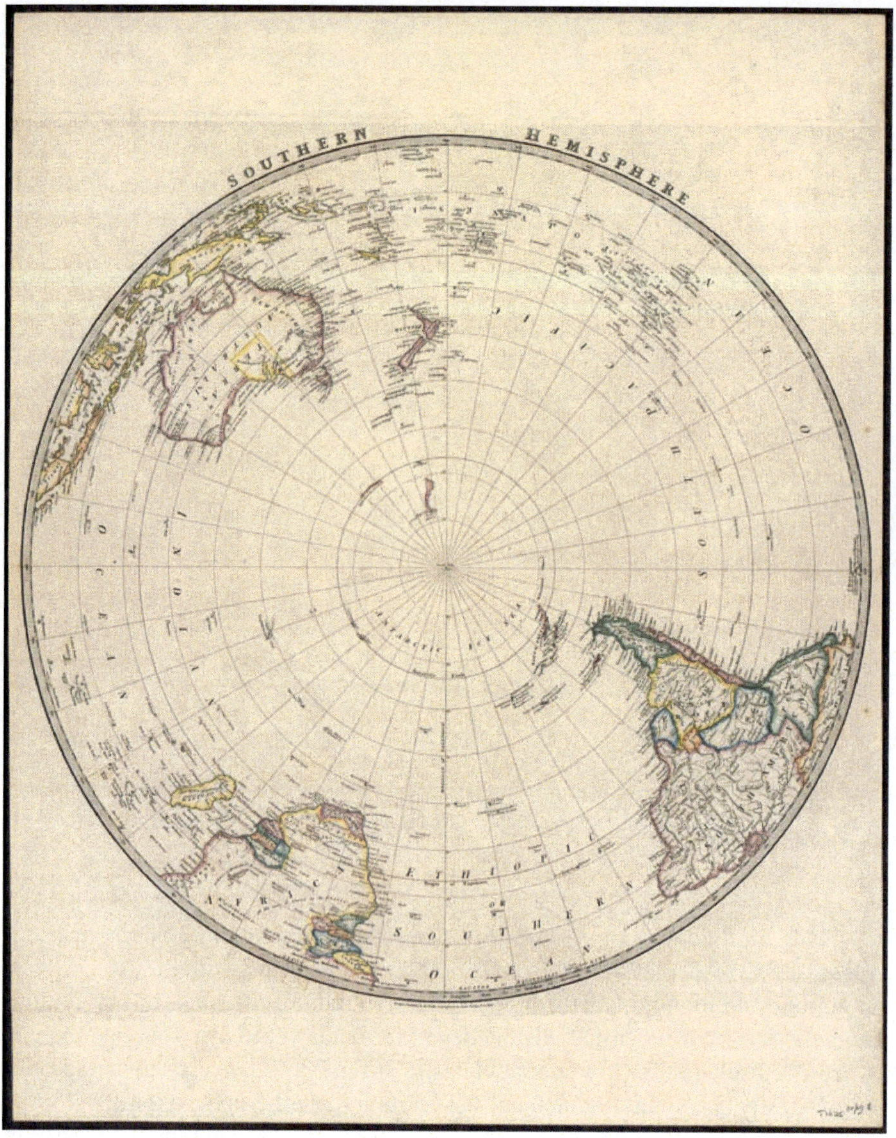

Fig. 5.17 Wyld *Southern hemisphere* (1849)

the imperial power. Scientific expeditions and private companies that operated in the southern seas broadened the registers of sub-Antarctic and Antarctic lands, enabling the constant update of official cartography. This, in turn, validated the information that fueled the atlas editorial business, a form of spreading geographical knowledge that contributed to place England as the imperial core of the world politics and commerce (Yan 2007).

5.5 Juxtaposed Territorialities

Geographical and cartographical works funded by Republican governments since the mid-19th century constitute a first stage in the construction of national territorial images, in which there is a clear difference between those territories effectively incorporated in state sovereignty and those claimed to be colonial inheritance, according to the 1810 *uti possidetis* interpretation. This rule, exclusive of American international law, intended to provide elements to solve two fundamental issues: the establishment of new state frontiers and the territorial ambitions of imperial powers. As regards the former, it suggested the adoption of the borders of Spanish administrative units upon independence and, facing the threat of colonization projects, it stated that all territories in Hispano-America, even those which had not been colonized, had formed part of Spanish domains and, thus, there were no lands not owned by anybody. Although the first international treaties considered it, this doctrine failed to be effective in preventing foreign interventions or territorial disputes, since the *res nullius* concept was present in the spatial conceptions. Only effective and permanent occupation granted actual dominion.

Until the conditions to the material incorporation of claimed territories were not met, the concepts of "interior frontier" and "desert" contributed to shape an imaginary involved in the construction of a national territorial project.

Although in the 19th century the concept of *frontier* linked to sovereignty was accepted, "the adjective "interior" refers to the vindication as their own of all geographical areas, even those located beyond the actual frontiers, within the inaccurate borders inherited from the Viceroyalty" (Quijada 2000: 378–379). Another recurring idea in the Río de la Plata political rhetoric was the idea of "desert", which in Patagonia represented, apart from the barren conditions of the plateau, the demographic emptiness, the dispute on the scarce resources and the unproductiveness; also "the unknown that maps should exorcize" (Lois 2018: 144–145). The ideological complex desert/savagery that, as Navarro Floria (2002) explains, constituted a cornerstone in the discourse of conquest, led to defining the inhabitants living beyond the frontier as an otherness that, from the imperatives of the state and capitalist order, should be tamed, or even suppressed. Floria notes the coexistence of various images of the desert in the official discourse of the second half of the 19th century. While the territories to the north of Negro River nurtured the utopia of progress as from the idea of a "desert to be transformed", Patagonia still produced ambivalent scenarios. Besides the traditional perception of a useless, infertile and uninhabitable territory, there appears another image boosted by the possible dispute with Chile, conceiving the region as a "fertile desert" such as La Pampa (Navarro Floria 2002:146–147).

Maps participated in the ideological construction of southern territories reinforcing the idea of a space sparsely peopled or inhabited by savages by including cartographical blanks and omitting any references to the indigenous territoriality.

C. Gay's maps of Chile outline the Patagonian cone, including a few toponyms on the coast and, in the interior, just the toponym Patagonia and the depiction of the Río Santa Cruz basin, explored by Fitz Roy. On the map of the Magellan Strait,

Gay adds the legend "unknown lands" and the toponyms "Western Patagonia" and "Eastern Patagonia".[70]

The *Mapa de la República Argentina* (1876) produced by A. Seelstrang and A. Tourmente for the R. Napp, aimed at promoting the country in Europe, anticipates the absence of interior frontiers. Cartographical blanks not only evidence geographical ignorance but also contributes to set the idea of vast lands available for agricultural production.[71] The book bolsters this view by stating that "almost half the territory is uninhabited" and, between brackets, adds: "(since we cannot call the savage and scarce tribes wandering across this huge territory "population")". This contemptuous attitude, barely differentiating indigenous people from nature, fails to prevent the awareness of their superiority, and that the key to their successful resistance lay on their extensive knowledge of the territory: "nobody knows the "desert" better, circumstance from which they take advantage with their shrewdness and peculiar movements".[72]

An important source to access to indigenous territoriality before the conquest is George Ch. Musters' account of his voyage around the Patagonian interior, between 1869 and 1870. Whether this former officer of the Royal Navy[73] made true the "so longed-for project of crossing the country from Punta Arenas to the Río Negro, Valdivia or Buenos Aires" (Musters 1979: 47) inspired on reading Darwin and Fitz Roy's accounts as he mentions in his chronicle, or else acted as an imperial agent, commissioner of an English land company or interested in the editorial business of voyage accounts, his detailed observations of the natural conditions for cattle raising proved to be very useful for the state territorial construction as well as for private economic interests. Colonists and investors, British in most cases, set their *estancias* on lands that were most suitable for the development of this activity, identified by Musters (Coronato 2016: 50). His book "At home with the Patagonians. A year's wandering over untrodden ground from the Straits of Magellan to the Río Negro", published in London in 1871, soon had a second edition in 1873. It was used as reference by all future explorers and his map of the territory "although incomplete, is not at least imaginary" (Muster 1979: 39). In fact, it is the first general map of Patagonia of the 19th century resulting from the observations in situ and the spatial knowledge transmitted by the tehuelches—Casimiro, Orkeke, Chiquichano, Foyel, Inacayal, Sayhueque and Hinchel—who guided him in the various stages of his itinerary (Bandieri 2005: 101). Elaborated on not specified European sources, with information obtained from his own experience and contrasting the data provided by his informants, Musters organized the map on a spatial-temporal sequence that describes by means of toponyms and dates an itinerary of over 1,500 miles (more than 2,700 km). From the indigenous approach of "lived space", it represents just the natural elements (hydrography and orography) and describes a historic route of the tehuelches by placing their whereabouts—sites frequently occupied due to the availability of water, pasture, wood and hunting spots. It also provides accurate references to the territory functioning: location of nations, use of space (areas and routes) and inter-ethnic relations. The view of the British subject is revealed when referring to the activity of "our missionaries" in the region, to the friendly treatment the natives had toward the foreigners, "mainly when these are English" and to the constant labor of maritime survey performed by the Admiralty ships. It also underlines the regular communications established between Port Stanley, Magellan and Patagonian coasts, which anticipate the conformation of a taxation economic space for the Empire.

Other foreigners were lured by the Patagonian territories that maps of mid-19th century depicted as an unknown country, independent of the new republics.[74] Orllie Antoine de Tounens' motivation was not curiosity or fascination for the experience of traveling to an unknown region but the possibility of making a political project come true. The French nobleman crossed the Bio Bio frontier in 1860 with the purpose of leading the Araucanians in the war against Chile and set up a constitutional monarchy. The self-proclaimed King of Araucania and Patagonia attempted to gain the favor of the main chieftains, let the Chilean and European press know about his endeavors and tried to gather support for his utopic kingdom.[75] Although Tounens' adventure came to an end in 1876 when he was captured and expelled to his country by Argentinian authorities, the unusual state project revealed the "structural foreignness" of Patagonia, an aspect of the regional reality that retrospective narratives that offer national historiographies fail to unveil. Contrarily, European and North American cartography do unveil it by representing "an alien geography, strange and separate" (Lois 2018: 138).

Except for a few foreign maps that incorporate southern territories into the new republics—for instance, the maps of South America of H. S. Tanner (1836, 1845), Mitchell (1846) and K. Johnson (1861),[76] French printer Pablo E. Coni was the one who produced the first identified maps that depict the whole of Patagonia as part of a state entity. He founded a printing house in Buenos Aires in 1863 devoted to scientific, cultural and educational publishing.[77] Maps published by this printing office differentiate the 14 provinces of the Argentine Republic and the Gran Chaco by means of colors ("Northern Indian Territory") and represent a vast jurisdiction of Buenos Aires that comprised the "Southern Indian Territory" and the "Patagonia". While the central map includes the territories up to about the 42° south latitude, following Peterman (1852) and Martin de Moussy (1865), the marginal map that depicts Patagonia and Tierra del Fuego is based on a cartographic source prior to the cartographical advances of Parker King and Fitz Roy. Thus, it draws the frontier with Chile at the Andes up to the Magellan Strait and fails to include the Malvinas/Falkland Islands.[78] By eliminating the boundaries of the effective exercise of state sovereignty in the indigenous territories—it just outlines the line of forts from the Salado River up to Bahía Blanca—and by identifying the different partialities, it offers a representation of the "interior frontiers" as a space of interaction and "formal representation of the control areas of the indigenous and "criolla" societies" (Mandrini 1992).

The integration of the southern territories up to the southern end—provided for in Sect. 5.2 of the Constitution of Buenos Aires State, passed in 1854 and in due force until the 1870 s[79]—expresses the territorial ambitions of the groups of economic and political power, firmly positioned opposite the rest of the Confederation after a decade of autonomy (1852–1861).[80] This claim by the Buenos Aires elite, or part of it, is understandable taking into account the interests originated by the advance of the cattle frontier in the humid pampas and the central role that Buenos Aires performed in this process during the government of Juan Manuel de Rosas, monopolizing the interethnic contacts and financing the policy toward the tribes (Quijada 2002). During the period 1853–1861, all initiatives referred to the economic incorporation with the aim of exploitation and colonization of Pampean-Patagonian territories under debate in Congress were exposed in both chambers of the Buenos Aires Legislative Branch and none in those of the Confederation (Navarro Floria 2002). Later, during the two decades following the political unification, during which Coni maps were produced, disputes over the political organization justified the intention of asserting territorial rights. On the one hand, the dispute between the national state and the provinces over the jurisdiction of the territories to be conquered, whose resolution for the former could be foreseen in Law 28 enacted in 1862, nationalizing the southern territories not claimed by the provinces and in Law 215 enacted in 1867 approving of the military occupation up to the northern margin of the Neuquén and Negro Rivers. On the other hand, the onset of the debate in the Senate of a bill on immigration and colonization promoted by President Nicolás Avellaneda, offering alternatives for economic diversification and suggesting other ways of using the land, apart from the traditional way: the *latifundium* (Navarro Floria 2002) (Fig. 5.16).[81]

From the perspective of territorial formation, cartographical representation of the Buenos Aires dominion over Patagonia supports the idea "that the provincial state would present (and to its successor in the autonomous 1852–1861) as the *foundation matrix* of the Argentine State as from Pavón" (Garavaglia 2007: 343). During the legislative debate of the 1854 Constitution, even when the impossibility of an immediate occupation was acknowledged, the necessity of demarcating the territory was grounded on the need to define the spatial scope of sovereignty and, in view of a future potential unification, on the convenience of protecting the territory, claiming that Buenos Aires should take it for the sake of the "Argentine association" (Aramburo 2016). The Buenos Aires government, as responsible for the Foreign Relations of the Argentine Confederation, had already assumed the defense of the purported rights over Patagonia when in 1833 it filed a claim against the English occupation on the Malvinas and when in 1847 it protested against the Chilean settlement on the northern coast of the Strait, on the understanding that it overrode the accepted natural boundary, the great mountain range of the Andes.[82]

Although Chile traditionally thought of the Andes as the eastern boundary of the country from Atacama to the southern end, as its Constitutions state, and the Argentine Confederation considered the Negro River as its southern boundary (Lacoste 2003), both countries aimed at incorporating Patagonia. Several documentary corpuses gave rise to different interpretations of the internal demarcations of the Spanish Empire upon independence. Thus, both governments performed jurisdictional acts, signed treaties with natives and promoted the first explorations to the interior of Patagonia, which started to unveil "its constitution, wealth, deficiencies and future" (Zeballos 1882, cited in Lois 2018: 129).[83] Meanwhile, cartographical production manifested territorial claims and consolidated the imaginary of the "desert" omitting any reference to the indigenous territoriality.

Chile's territorial expansion started with the occupation of the northern margin of the Strait of Magellan in 1843, and in 1852 a law was created in the Province of Arauco, which included the territories between the Bio Bio and the province of Valdivia. The establishment of the state hegemony over indigenous territories in the center-south of Chile resulted from the foreign colonization, favored since the 1840 s by statutory regulations that regulated the transfer of indigenous lands to private hands, and also from the territorial conquest started in 1862 with the gradual frontier advance, that came to an end with the military occupation between 1881 and 1883. Besides, the project of peopling the southern end contemplated the benefits that could bring a colony on the edge of the interoceanic channel, revalued due to the development of steam navigation. The first crossing of the channel by the steam ships of the Pacific Steam Navigation Company in 1840 inaugurated the route linking England to Valparaíso and El Callao Ports. This event marked the beginning of a new period in which the Magellan region definitely inserted into the international market. In 1843, a fort was founded on Santa Ana promontory, near the ephemeral settlement of San Felipe (1584), and between 1847 and 1848, Fuerte Bulnes settlement was moved 50 about kilometers to the northeast, to the site J. Byron named Sandy Point (1784), next to Carbón River. The colony started as a criminal-military establishment. Only in the 1870 s, Punta Arenas started to consolidate as a center of colonization due to the beginning of regular commercial navigation through the Strait, seal hunting, feathers and furs trading with natives, logging, gold and coal mining. Since 1877, sheep breeding—an activity that grew rapidly in the ecosystem of the steppes and in only two decades—consolidated in the regional economy (Martinic B. 1992, 2001; Inostoza Pino 2012).

C. Gay's maps of Chile, used by state and private agents until the 1870 s, anticipated the territorial occupation by omitting all references to indigenous territoriality in the Araucania and in the territories of southern Patagonia. Both Gay's map of the "Magellan Strait"—possibly elaborated prior to the peopling plan, since it indicates Puerto del Hambre on the site where Fuerte Bulnes[84] was settled—and the "Map of Chile" produced by D. Barros Arana by the end of the 1860 s,[85] represent the territories to the south of the 45° as an empty space with no demarcations. However, both suggest a territorial differentiation through main toponyms. The former uses the names "Western Patagonia" and "Eastern Patagonia" and the latter "Province of Llanquihue" and "Territory of Magellan" on the Pacific Ocean slope and "Patagonia" on the Atlantic Ocean slope. The omission of the Andes proves the lack of surveys and accurate topographical data and the uncertainty of the frontier with Argentina (Figs. 5.17 and 5.18).

Barros Arana, who acted as a diplomatic agent in Buenos Aires during the most critical period of negotiations—between 1876 and 1878—expressed on his map the most ambitious claim over the territories to the east of the Andes by representing almost all Patagonia, coinciding with the jurisdiction assigned to Chile upon Pedro de Valdivia's request by the mid-16th century. That is, a strip of 100 leagues from the coast of the Pacific toward the interior and from parallel 41° up to Magellan Strait. In the book "*Elementos de Geografía Física*" (1871) he clearly stated the territorial assertion by explaining that "to the south of the Patagonian Negro River, Chile believes to have titles of dominion although controverted by the Argentine Confederation" (cited in Lacoste 2003: 213).

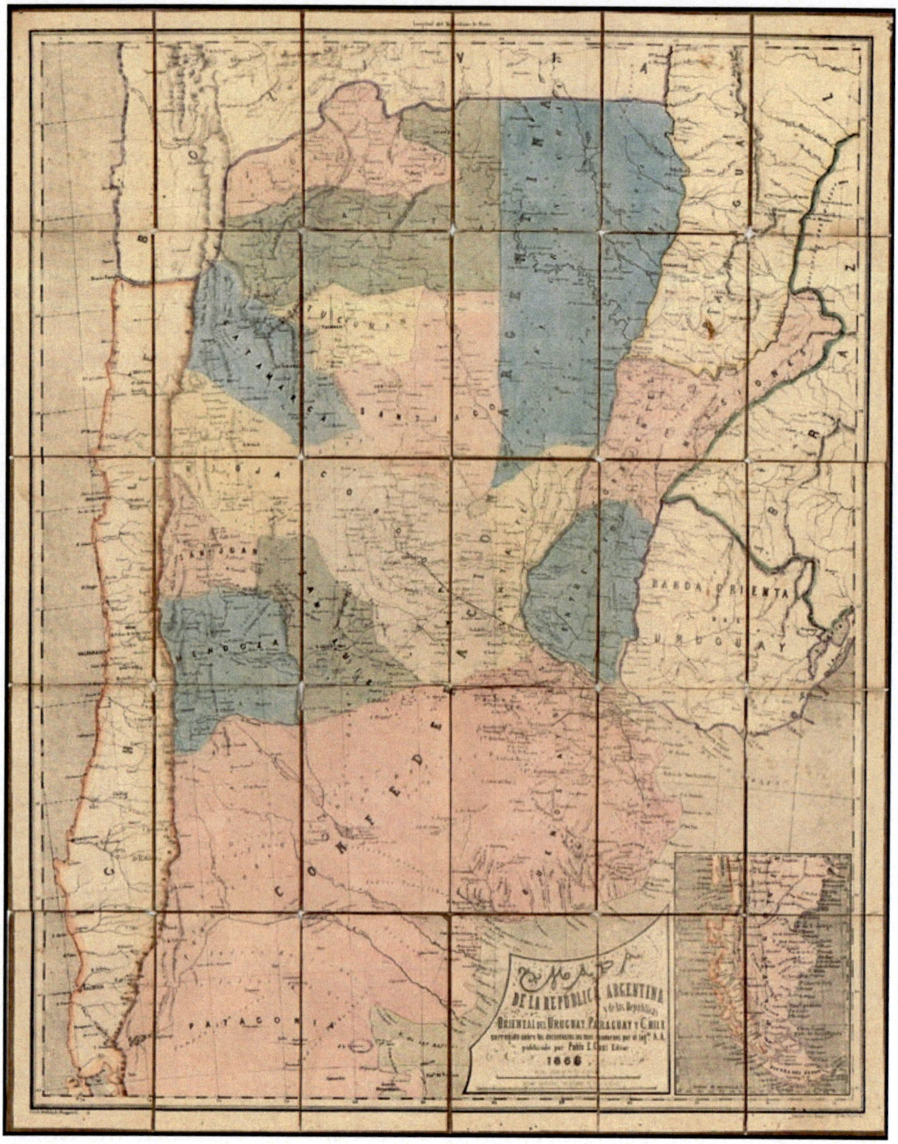

Fig. 5.18 *Coni Mapa de la República Argentina (1868)*

The aforementioned *Mapa de la República Argentina* (1875) produced by A. von Seelstrang and A. Tourmente is the first official map that offers the territorial image of the unified State-Nation, "unification that included Buenos Aires and Patagonia" (González Bollo 1998). Rather than being a scientific description of the territory, the map is a proposal of territorialization. Like the work in which it is contained explains, international boundaries—as well as inter-provincial ones—"are not defined accurately", and jurisdictional demarcations were based on a report of the Senate Commission in charge of the examining the bill for the creation of National Territories in 1871.[86] Territorial claims were indicated by identifying the occupied sites to the south of the Negro River due to land concessions and authorizations to colonize and exploit natural resources. Luis Piedrabuena's establishment on Pavón Island, at the mouth of the Santa Cruz River (1859), and on the Staaten Island (1862), the Welsh colony in Chubut (1865), Rouquaud colony in the lower valley of the Santa Cruz River (1872–1874) and, in the Magellan Strait, Tehuelche whereabouts that had been deemed to be the settlement of an indigenous colony. The Malvinas/Falkland Islands are depicted with double toponymy, English and Spanish, and there is no reference to Port Stanley, capital city of the British colony since 1845. The insular end is incorporated into the national territory, from the geographical conception of belonging of the Patagonia, even when Tierra del Fuego is considered as "a secondary region as regards its practical importance", as Napp explained in his work (Fig. 5.19).

While looking for an agreement to set the shared frontier, both countries organized the military advance over indigenous territories as a condition of effective and productive occupation. Following the so-called "Pacification of the Araucania" or "Conquest of the Desert", cartography—together with other state's tools as the census statistics and land registries—contributed to keep the indigenous people invisible (Otero 1997-1998; González Bollo 1999; Yuln 2014). In Chile, the Mapuches were assimilated as an ethnic minority and subject to a process of marginalization and impoverishment, product of a settling policy that isolated and concentrated people in reductions or reserves that represented just a 6% of Mapuche historic territory (Boccara and Seguel-Boccara 1999). In Argentina, autochthonous peoples were made citizens, acknowledging their rights as primarily pertaining to their national territory, and the lands that belonged to the communities in the Pampean Region and the north of Patagonia were transferred to the State. Despite of the fact that legislation provided for the creation of agricultural-pastoral colonies of mixed nature, it was more frequent that indigenous groups settled after delivery of lands of various extensions to the caciques: allies gained ownership title while most of them were granted usufruct rights to the lands (Quijada 1999, 2000; Bandieri 2005).

Likewise, the system of settlements through missions was conceived, which was coordinated by the Salesian Order, congregation founded in Turin in 1859 by Catholic Priest G. Bosco, and counted with an ecclesiastic jurisdiction conferred by the Holy See in 1883 (Nicoletti 2008). The Vicariate of Northern Patagonia, based in Carmen de Patagones, and the apostolic subprefecture in northern Patagonia and Tierra del Fuego, based in Punta Arenas, formed a transnational missionary territory that comprised the south of Chile and Argentina and the Malvinas Islands. The Order's cartography represents it as a space with no political demarcations, in which there appear the Salesians and the Daughters of Mary Help of Christians' itineraries in their itinerant missions in northern Patagonia. It also depicts the reductions in Tierra del Fuego: San Rafael on Dawson Island, granted by Chilean government in 1889 for 20 years and Nuestra Señora de la Candelaria, founded in the north of the Isla Grande de Tierra del Fuego in 1893.[87]

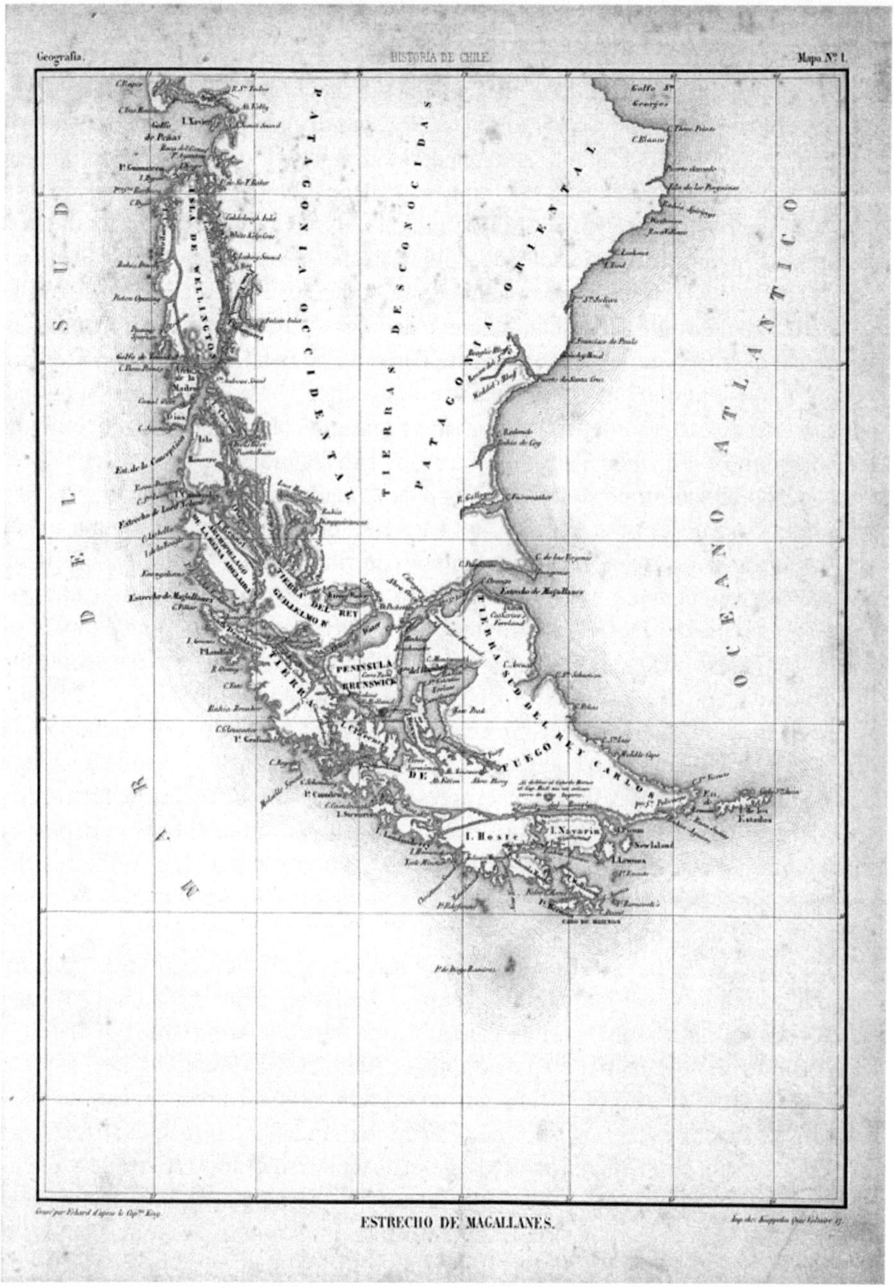

Fig. 5.19 *Gay Estrecho de Magallanes* (1854)

The Boundary Treaty of 1881 was the first official step toward the definition of the territorial jurisdictions in the regions under dispute: Patagonia itself, Magellan region, Isla Grande de Tierra del Fuego and islands of the archipelago located to its south, south-west and west. The map published by the Chilean government in August 1881 represents the Argentine and Chilean demarcation proposals (May 1879 and June 1881, respectively) as well as the boundary agreed under the treaty, signed on 23 July 1881. The map shows that the Magellan steppe, to both sides of the Strait, represented a key area within the diplomatic debate. This occurred mainly after the first Chilean expedition to the interior of Tierra del Fuego, led by Ramón Serrano Montaner, at the beginning of 1879, that confirmed that the natural pastureland in the center and north of the Isla Grande were suitable for sheep breeding (Fig. 5.20).[88]

Omissions and inaccuracies of the boundary treaty, and mainly, discrepancies over the interpretation of some of its terms, gave birth to new controversies during the demarcation tasks on the southern Andes, where the line over the highest summits of the Andes fail to coincide with that over the watersheds. The "*Carta de la Extremidad Meridional de Sud América*" (1889), elaborated under the direction of the expert D. Barros Arana, represents the two tracing alternatives: the orographic, promoted by Argentina, and the hydrographic, claimed by Chile.[89] In 1898, the issue was submitted to the British Crown arbitration, whose arbitral award, released in November 1902 in the tense context of war preparations, set an ad hoc boundary that divided the territory under dispute in a more or less equitable fashion (Lois 2010).

Whereas the Patagonian space was distributed among the National Territories created in Chile and Argentina[90] and legally confirmed by successive bilateral agreements, the controversy over the sovereignty on Malvinas Islands—"left aside, as a frozen asset within the diplomatic negotiations" (Ferns 1984: 237)—maintained an overlapping of frontiers in the South Atlantic Ocean. The map of the "*Gobernación de la Tierra del Fuego y las Islas Malvinas*", elaborated by M. F. Paz Soldán (1888),[91] reveals the territorial claim by depicting the Malvinas Islands in an inset, on a smaller scale, in the center of the map. The map of Tierra del Fuego and the Malvinas Islands, elaborated by A. von Seelstrang in 1885 and included in the "*Atlas del Instituto Geográfico Argentino*" published in 1893, presents the same design and adds: "*Las Malvinas hánse [sic] agregado a este mapa por los antiguos y bien fundados derechos que tiene la República sobre ellas*" (Lois 2007: 128) (Fig. 5.21).

The "dismal, desolated and windy" South Atlantic Ocean archipelago acquired a definite identity within the commercial and maritime network of the British Empire during the second half of the 19th century (Warnick 2008) upon consolidating as a strategic enclave for the dominion of the South Atlantic Ocean and the control of the traffic in the Pacific and Indian oceans, as important as Gibraltar, Malta, Aden and Hong Kong, as declared in 1853 memorandum (Moreno Alonso 1983). The Malvinas/Falkland Islands became the core of the maritime and insular South Atlantic space targeted by Great Britain's claims, territorialized through cartography as one of the circumpolar regions of imperial interest and influence in the southern hemisphere.[92]

After receiving the official status of colony in 1843, during the first years of civil administration, the inconsistent policy persisted, until the following decade in

Fig. 5.20 Barros Arana *Mapa de Chile* (1872)

Fig. 5.21 Seelstrang and Tourmente *Mapa de la República Argentina* (1875)

which the possibilities offered by its strategic location were taken into account in the development of private projects. The Falkland Islands Company was founded by Samuel Lafone, a British merchant based in Montevideo who, on the basis of information provided by L. Vernet, presented a proposal of colonization and, although never visited the islands, became the most prominent investor, setting an example for the absent businessmen (Warnick 2008). In 1846, he received the concession of most of the southern half of Eastern Malvina/East Falkland, fishing rights and wild cattle handling. Between 1851 and 1852, with the support of investors from London, it was granted the Royal Charter. The Royal Falkland Land, Cattle, Seal and Fishery Company was the most important land owner, monopolized transport and trading, and took part in local government councils. Thus, it concentrated local political and economic power (Harambour R. 2016). It also gained control over ships repairs and provisioning, profiting from the intensification of the traffic to and from the Pacific

Ocean through Cape Horn, movement fueled by the exploitation of guano in Peru and the discovery of gold in California and Australia (Warnick 2008). By the mid-1850 s, the depletion of wild cattle accelerated the shift to sheep breeding and the company got involved in the transport of sheep and workforce from Great Britain and rapidly increased the activity, whose surplus enabled the shipping of the first cattle stocks to Magellan, Santa Cruz and Tierra del Fuego since 1876 and during the 1880s.

While the Malvinas/Falkland were the United Kingdom's possession, the southern territories in Chile and Argentina were part of the imperial periphery, since the alliance among British and national states capitals gave place to the integration of the southern region into the imperial circuits as a territory specialized in sheep breeding (Harambour R. 2015).[93] State policies facilitated the access to the land through the sale or lease of vast extensions, special concessions and transfers (Belfiori 1977; Barbería 1991; Martinic B. 1992) and the flow of English investments favored the formation of powerful livestock enterprises that monopolized the best grazing areas. Apart from the production, they controlled transport and trading of wool and meat, credit, industries related to livestock and construction activities and commerce in general, all through Chilean and Argentinean Patagonia (Harambour R. 2017).

In the adaptation process of sheep breeding to the possibilities and restrictions of the steppe ecosystem and the organization of the activity according to the British productive model, the Magellan Region—based in the city-port of Punta Arenas—together with the rest of Patagonia and Tierra del Fuego, "adopted the virtual character of colonial tax payers" (Martinic 1992: 797). The spatial framing of Colton's map "Southern Portion of the Peninsula of Patagonia" (1879),[94] based on Seelstrang's map (1875), coincides with the area of expansion—Malvinas and Magellan herbaceous steppes-band anticipates an image of the cattle territory as a space with no frontiers, "a kind of plain surface on which sheep and flock masters wander with no barriers" (Bascopé 2008: 25) (Fig. 5.22).

Fig. 5.22 Prieto *Carta que demuestra las diversas proposiciones de arreglo de la cuestión de límites chileno-argentina* (1881)

Fig. 5.23 Paz Soldán *Gobernación de la Tierra del Fuego y las Islas Malvinas* (1888)

Fig. 5.24 Colton *Southern Portion of the Peninsula of Patagonia* (1879)

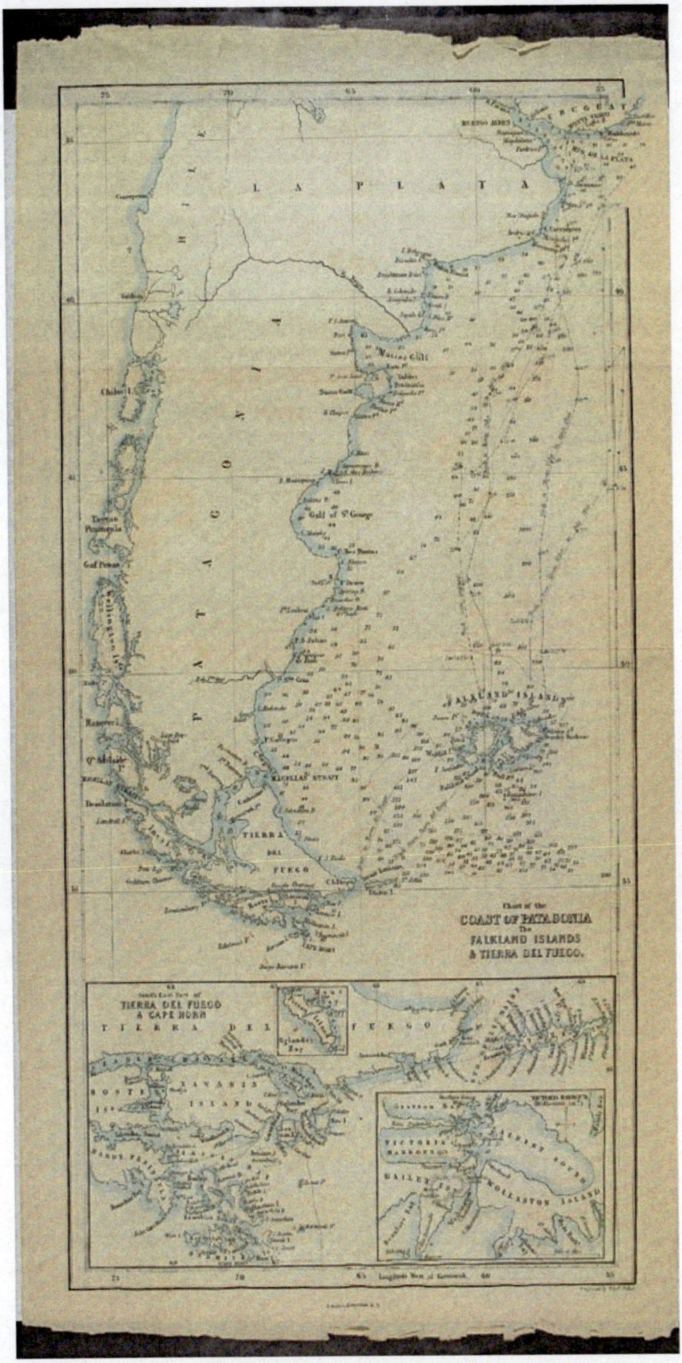

Fig. 5.25 Parker Snow *Chart of the Coast of Patagonia, the Falkland Islands & Tierra del Fuego* (1857)

Fig. 5.26 Martial *Carte Ethnographique* (1883)

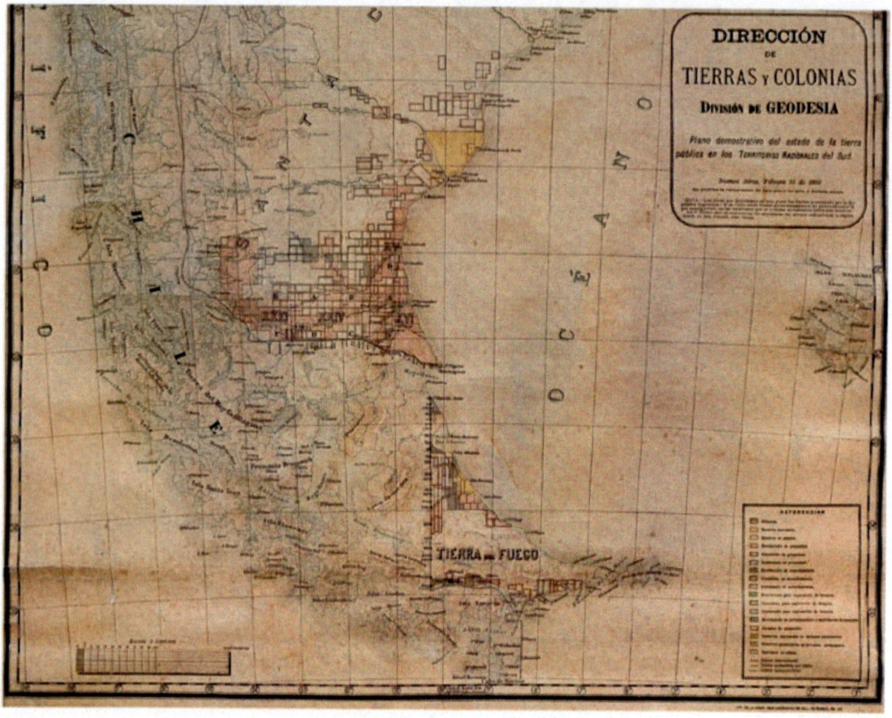

Fig. 5.27 *Plano demostrativo del estado de la tierra pública en los Territorios Nacionales del Sud* (1900)

Unlike northern Araucania and Patagonia, where military operations and territorial policies managed to disrupt the indigenous territories expeditiously, natives' resistance conditioned cattle colonization of Magellan, Santa Cruz and Tierra del Fuego for over a decade. Explorers, missionaries and travelers who moved around the southern region after the boundary agreement between Argentina and Chile confirmed that the indigenous people kept living their own lifestyle, using their hunting areas and whereabouts. They also observed growing inter-ethnic conflicts, resulting from the colonialist advance, particularly during the last decade of the 19th century (Martinic B. 1979, Harambour R. 2015). Relations with northern tehuelches were typically peaceful, whereas contacts with canoers from the western area of the channels in Magellan and Tierra del Fuego and with hunter groups from the center and north of Tierra del Fuego had an inimical nature due to the indigenous resistance strategies and the miners, sealers and *estancias* and the abuses of cattle companies' employees. Victims of different forms of violence, of the weakness of state institutions and public officers' indifference, the Alacalufes and the Selknam were cornered in a small area of their territory or moved to Salesian missions, where if they survived the infectious diseases epidemics were transformed into Christian workers (Bascopé 2010, 2011). These centers that, during the first years, accommodated a significant number of people—San Rafael mission held up to 550 natives and La Candelaria over one hundred—toward 1910 were almost deserted, a situation that led to the creation of missionary stations in the interior of Isla Grande, Santa Inés or Río Fuego and San José at the source of Fagnano Lake.

The Yagan territory was the field of action of the South American Missionary Society, an organization based in London that, like the Salesian Society, developed its missionary project with broad autonomous organizational powers. In 1869, the superintendent of Keppel mission was appointed Bishop of an Anglican Church diocese with jurisdiction over the Malvinas/Falkland and Tierra del Fuego. From the estancia-mission settled in 1854 on the islet of Malvinas/Falkland, authorized by the Foreign Office a year later, regular voyages to Tierra del Fuego were organized to temporarily transport natives who were instructed in Christianity (Bascopé 2016). The interaction experience with people from Tierra del Fuego and the fact of knowing their language enabled the evangelization in the indigenous territory itself and, since 1871, Ushuaia, on the northern coast of the Beagle Channel, became the center of the missionaries' permanent residence. The Admiralty Hydrographical Office published in 1877 Fitz Roy's map of the south-east of Tierra del Fuego, adding a detailed sketch of the bay that shows the mission site and the protected port of easy access.[95] The proximity of the population center around the Argentine Naval sub-prefecture established in 1884 altered the functioning of the mission and the epidemics caused a rapid and significant decline in the native population, which forced its closing some years later (Bridges 1978). W. Parker Snow's map, commander of the schooner of the mission during the first voyages to Tierra del Fuego, represents the Anglican missionary territory indicating the maritime connection between Keppel Island and the channels of Tierra del Fuego and the missionaries' itineraries up to the insular southern end, where the Missionary Society kept its last stations: Wollaston on Bayly

Island (1888–1892), Tekenica on Hoste Island (1892–1906) and Douglas River on Navarino Island (1906–1917) (Martinic B. 2009) (Fig. 5.23).[96]

Fitzroy, Robert. The south-eastern part of Tierra del Fuego with Staten Island, Cape Horn and Diego Ramirez Islands. Hydrographic Office. 1877, 1st publ. 1841
http://collections.rmg.co.uk/collections/objects/540852.html

Examining both ethnographic maps of the *Mission Scientifique du Cap Horn* (1882–1883)[97] and the *Expedición Austral Argentina* (1883)[98] that depict native population distribution at the initial stage of sheep expansion, and cadastral maps that show the configuration of territorial ownership at the onset of the 20th century in Chile[99] and in Argentina[100], we can visualize the conformation of cattle territory and the natives' withdrawal and territorial loss. State cartography accounts for the private appropriation of the Magellan-Fuegian space, presenting the division and distribution of expropriated lands, transformed into public land, as a way of making national sovereignty effective over southern frontiers. While southern Tehuelches and the Selknam were territorially reduced throughout the first decades of the sheep cycle boom, canoer peoples, Yagans and Alacalufes—geopolitically marginal as regards cattle industry—experienced a slower withdrawal and kept key sites of their territories until well into the 20th century (Harambour R. and Barrena Ruiz 2019: 27; Serrano 2006).

5.6 Conclusions

Cartographical analysis has enabled us to think of territorial construction of the southern end of America at different moments of its introduction to the world market and during the processes of imperial and national territorialization and indigenous deterritorialization.

Cartography becomes an effective discursive and visual device to express territorial projects and claims, and contributes, in turn, to validate the legal-political arguments used to legitimate them—discovery doctrine, effective occupation principles, *terra nullius* and *uti possidetis*. It also upholds the concepts or core ideas that support territorial images and geopolitical fictions.

European and American maps produced after the disintegration of the Spanish Empire reveal the political and economic interests underlying the design of colonialist plans. These maps ignore independent government territorial claims, based on the alleged colonial inheritance, and depict Patagonia and Tierra del Fuego as differentiated entities, out of any state control, inhabited by free peoples, pagan and politically fragmented.

The idea of the frontiers of the new republics between the 44° and 41° south latitudes acknowledges a space controlled by politically and territorially sovereign nations, more developed than hunting peoples and gatherers who inhabited the south of the river corridor Limay-Negro. Such a hierarchical differentiation between nations and territories, cartographically manifested in the division of Patagonia by using two distinct scales and in the legends that lead to ascribe the southern territories under the category of *res nullius*, would account for the integration of Araucania, the north of Patagonia and the *pampas* to the new republics before effective occupation. Demarcation is in no way arbitrary, although A. d'Orbigny failed to acknowledge the reason of its adoption and reproduction on maps, since from this influential naturalist's point of view **all free peoples** "from the 34 degrees south latitude up to the end", beyond their civilization development, belonged to the same "**savage type**", that is, communities that were naturally unadaptable to western culture (Navarro Floria 2005). Nothing results more obvious than the ideological compromise underlying a territorial configuration set toward the mid-19th century and persisting in the first national cartographies as "interior frontiers".

Thus, imperial and national cartographies took part in the ideological construction of the southern territories, describing the inhabitants as barbarian and savage or else omitting on the maps any references to indigenous territoriality and recreating the "desert" notion. These strategies aimed at showing an unknown space as "the promise of all possibilities" (Silvestri 2011: 97), at justifying integration policies—the military conquest of the northern territories and the private conquest of the southern territories—and, lastly, at making invisible the indigenous population incorporated into the citizenship.

Sources belonging to the first decades of the 19th century also enabled the valorization of the South Atlantic Ocean from specific strategic and economic interests. While the interoceanic traffic intensified due to the commercial possibilities in different areas of the Pacific basin, the coasts of southern Patagonia and Tierra del Fuego, the sub-Antarctic islands and the coasts known as Antarctic Peninsula attracted businessmen related to the hunting of sea mammals. This activity constituted the first exploitation cycle in the southern region, which started in the last decades of the 18th century with an intensive exploitation of the resource and reached the highest profits in the oil and skin trading in the 1820 s and persisted with less profits for the second half of the 19th century.

Great Britain took the lead in the exploration of the insular end of the continent and the islands to the Pole and in the cartographical production, due to official and private expeditions. Thus, it broadened its knowledge of ports of call and bases for hunting activities and identified a space of growing geopolitical and economic interest focused on the interoceanic routes that include the hunting areas of pinnipeds and cetaceans up to the 60° south latitude. Toponymy fulfills the territorializing function through a process of re-naming, strategy that substitutes Spanish, Dutch and French names and inaugurates a process of loss of information regarding indigenous spatiality. Geopolitical intentionality is also expressed by representing a circumpolar territoriality that reveals the British positioning in the southern hemisphere and the control of the possible accesses to the Pole from imperial possessions in the Atlantic and the South Pacific and the Indian oceans.

The first republican jurisdictional acts in the southern region targeted strategic sites, contemplated in the European expansive plans. On the one hand, the Political and Military Command created by the government of Buenos Aires in 1829 established on the Malvinas/Falkland Islands to control foreign hunting activities and to start colonization. However, the project was interrupted by the British occupation in 1833, a decision that was forced by the increase of international competition for the use of maritime resources in the South Atlantic Ocean. On the other hand, the Chilean settlement on the Magellan Strait, started in 1843, settled around the penal colony of Punta Arenas and generated the conditions for a new exploitation cycle, initially associated to coal and gold mining and, with the introduction of sheep by the end of the 1870s, to extensive cattle-raising. With state policies that favored the access to land and enough funds, predominantly English, breeding extended during the 80s to Santa Cruz and Tierra del Fuego and transformed the northern and southern steppes into a stockbreeding territory linked with the world's market through imperial networks. Bearing in mind British significant commercial and financial participation in the national economies expansion during the second half of the 19th century, it is understandable that the controversy over the Malvinas/Falkland Islands became an accepted element in Anglo-Argentine relations and that the international arbitration that put an end to the dispute over the Argentinean-Chilean boundary in Patagonia was taken over by the British Crown.

Considering the demarcation criteria used in state cartography with diplomatic or propagandistic purposes—the geopolitical, that conceives the boundaries as a separation line between two states, and the environmental, based on the formulation of natural regions, we observe some features of the cartographic representation of the southern territories of the 19th century. On the one hand, bearing in mind that the state territoriality only starts to be effective during the first decades of the 20th century, the lack of correspondence between jurisdictional limits and the spaces effectively ruled by the States becomes evident. On the other hand, geographical ignorance conditioned the representation and justified the omission in tracing the Andes to the south of the 45° south latitude in some maps, and in the case of the Magellan steppe, the difficulty in defining accurately a "natural frontier". However, the image of an empty space was functional to the state interests in the promotion of the region. From the assumption that "the territory is the natural base of the State" (Quintero 1999), cartography serves the purpose and delimitates intra and inter-state jurisdictions and provides cadastral maps to transfer indigenous lands to private hands. Meanwhile, cattle expansion acknowledges the end of the Magellan steppe as a "natural frontier" and ignores the geodesic boundary that gives place to the geopolitical fiction of the frontier as a rim.

While cartography rehearsed demarcation criteria in the abstract space, incorporating vast extensions considered *res nullius* into the state sovereignty, socio-spatial dynamics were defined by territorial actors' strategies—independent indigenous people, private concessionaires of the conquest: *estancieros* (ranch owners) and missionaries, and state agents under their service—and extra-territorial ones—capitalists related to trading, transport and finance. State territorialities and, thus, frontiers, resulted from later interventions, policies and control mechanisms implemented during the first decades of the 20th century within a context of reconfiguration of strength relations (Figs. 5.25, 5.26, 5.27 and 5.8).

The study of maps, from an "analytical approach of space representations", that is, an analysis of the power relations that shape the territory and that are mainly expressed in territorialities, more or less ephemeral or long-lasting, opens different reading levels. The interpretation results from both the inquiring into what is implicit—Harley's "hidden agenda" (2005)—and the **difficult comprehension of what is apparent**, an operation that aims at giving sense to what is visible and largely ignored. As any other source, maps veil and, at the same time, unveil. The result depends on the questions and proceedings in the search of sense, as well as on the sense of the search.

Notes

1. The Requirement was a declaration, read aloud to the natives before attacking them, under which they were forced to subject themselves to the Spanish and to become Christian. Although native inhabitants were legally free, those who were under the *encomienda* system lived within a virtual slavery, since they were not able to exercise their freedom by deciding not to work for the Spaniards (Pagden 1988). The New Laws of 1542 abolished the encomienda system due to abuses and because it was a threat to the authority of the Crown and Church, though in practice, its dismantling was slow and hard to accomplish.
2. The hospitality and gifts practices, including the provision of material goods, prestige, objects, honors and titles, favored the concentration of power in the socio-political indigenous system and constituted a significant part of the reciprocity relationships kept by the indigenous sociability and authority modes, both intra- and inter-tribes (Quijada 2002: 118–119).
3. J. de la C. Cano y Olmedilla, *Mapa Geográfico de América Meridional*, second edition, 1776. National Historical Archive, Madrid, Signatures 676 and 738, 321 and 328. Repository: David Rumsey Map Collection, David Rumsey Map Center, Stanford Libraries.
4. M. Lastarria *Mapa Geográfico de América Meridional* y *Carta Corográfica del Virreinato de las Provincias del Río de la Plata*, 1804, *Colonias orientales del río Paraguay o de la Plata.* Documents for Argentine History, Philosophy School, Buenos Aires, 1914. Vol. III. Digital repository: Robarts Library, University of Toronto.
5. According to some authors, the scheme of alliances and parliaments developed by the Kingdom of Chile as from the second half of the 17th century and adopted during the 18th century by Buenos Aires, Córdoba and Mendoza governments could be deemed as an indirect way of possession of the territories that made it possible to assure Spanish sovereignty upon other powers' possible claims (Lacoste 2003: 180 and ff.).

6. W. Woodbridge, *Moral & Political Chart of World. Exhibiting the Prevailing Religion, Government degree of Civilization and Number of Inhabitants of each Country*, 1828. Repository: David Rumsey Map Collection, David Rumsey Map Center, Stanford Libraries.
7. Naturalist Alcide d'Orbigny was sent to South America by the National Museum of Natural History in Paris, as recommended by his teacher, the anatomist Cuvier, and the outcome of his labor was published in the monumental piece *Viaje por la América meridional (el Brasil, la República Oriental del Uruguay, la República Argentina, la Patagonia, la República de Chile, la República de Bolivia, la República del Perú) efectuado durante los años 1826, 1827, 1828, 1829, 1830, 1831, 1832 y 1833,* published in Paris and Strasbourg between 1835 and 1847, In 1939 he published *El hombre americano*, dedicated to A. Humboldt.
8. W. Woodbridge, *A system of Universal Geography*, 1831. Repository: Boston Public Library.
9. A. Balbi, *Novísima Geografía Universal*, 1848. Repository: Library of Monserrat Abbey - Digital
10. A. Brue, *Nuevo Mapa de las Repúblicas del Río de la Plata, Chile, Paraguay, Uruguay y parte de los países vecinos*, 1843. Repository: National Library of Uruguay. Digital Collections.
11. R. Wilkinson, *South America*, 1806. Repository: David Rumsey Map Collection.
12. J. Cary, *South America south*, 1807. Repository: David Rumsey Map Collection.
13. L. Delarochette, *Colombia Prima or South America*, 1811. Repository: David Rumsey Map Collection.
14. A. Brue, *Amerique Meridionale*, 1816. Repository: David Rumsey Map Collection.
15. A. Brue, *Carte Generale de l'Amerique Meridionale et des isles qui en dependent*, 1826. Repository: David Rumsey Map Collection.
16. A. Brue, *Carte Generale de l'Amerique Meridionale et des isles qui en dependent* 1839. Repository: David Rumsey Map Collection.
17. C. Malte Brun, *Chili, Plata, et Patagonie*, 1837. Repository: David Rumsey Map Collection.
18. A. Brue, *Carte Generale de l'Amerique Meridionale et des isles qui en dependent*, 1826. Repository: David Rumsey l Map Collection.
19. H. S. Tanner, *Chile, La Plata and Montevideo*, in *Tanner's Universal Atlas*, 1836. Repository: David Rumsey Map Collection.
20. T. Bradford, *A comprehensive atlas, geographical, historical & commercial*, 1835. Repository: Boston Public Library, Norman B. Leventhal Map Center.
21. G. W. Colton, *South America (political)*, 1858. Repository: David Rumsey l Map Collection.
22. J. Arrowsmith, *The Province of La Plata, the Banda Oriental del Uruguay & Chile, chiefly from M.S. document communicated by Sir Woodbine Parish,*

1839. Repository: National Library of France, Charts and Maps Department, Digital.
23. Lois (2015) notes the cartographical association of the red color to the British Empire.
24. A. Peterman, *The provinces of the Río de la Plata and adjacent countries: chiefly from the map constructed by J. Arrowsmith for Sir Woodbine Parish, K.C.H., drawn by Augustus Peterman,* 1852. Repository: The Huntington Library, San Marino, California. Digital.
25. C. Gay, "*Mapa para la inteligencia de la Historia física y política de Chile*", *Atlas de la Historia física y política,* 1854. Repository: Missouri Botanical Garden—Digital.
26. V. Martin de Moussy, *Atlas de la Confédération Argentine,* París, 1865. Repository: Spain National Library –Hispanic Digital Library.
27. V. Martin de Moussy, *Carte du territoire indien du sud et de la región des pampas,* 1865 Repository: Spain National Library –Hispanic Digital Library.
28. V. Martin de Moussy, *Carte de la Patagonie et des archipels de la Terre de Feu des Malounies et des còtes occidentales*, 1865. Repository: Spain National Library –Hispanic Digital Library.
29. V. Martin de Moussy, *Description* […], 1860. T. 1, pp.70-71. Repository: Spain National Library –Hispanic Digital Library.
30. E. Bransfield, *Chart showing the discoveries and approximate track of Edward Bransfield, Master R.N. in the hired brig "Williams" of Blyth, when exploring the South Shetland Islands and the northern extremity of Graham Land in the year 1820, 1822.* Repository: National Library of Australia. G. Powell, *Chart of South Shetland including Coronation Island & c. from the exploration of the Sloop Dove, in years 1821 and 1822,* 1822. Repository: State Library New South Wales.
31. J. Weddell, *The Polar part of the Southern Hemisphere, Exhibition a comparative view of land and sea with that of the Northern,* 1825, Repository: Linda Hall Library, Digital Collections.
32. J. Weddell, *Chart of the tracks of the Vessels Jane & Beaufoy in their Southern voyages of investigation in 1820, 1821, 1822, 1823 & 1824 by J. Weddell R.N.*, 1825. Repository: Linda Hall Library, Digital Collections.
33. J. Weddell, "*Chart of South Shetland, constructed from observations made during three voyages in the years 1820, 21, 22 & 23. To these islands, by J. Weddell, RN*", 1825. Repository: Linda Hall Library, Digital Collections.
34. W. Smith, *The Voyages of Captain James Cook*, Vol. 1, 1842, British Library - Digital.
35. J. A. Colnett, *A voyage to the South Atlantic and Round Cape Horn into the Pacific Ocean: for the purpose of extending the spermaceti whale fisheries and other objects of commerce* […], London, 1798. Repository: University of Alberta Libraries—Digital.
36. Vargas Ponce, *Relación del último viaje al estrecho de Magallanes de la fragata de S.M. Santa María de la Cabeza en los años de 1785 y 1786 […],* 1788. Repository: Library of Seville University-Digital.

37. J. Cook, *A Chart of the Southern Extremity of America*, 1777. Repository: David Rumsey Map Collection. Image in Chapter 3.
38. Selma, *Carta esférica de la parte sur de la América Meridiona*l, 1788. Repository: John Carter Brown Map Collection.
39. W. Faden, *Chart of the Coasts of South American from Río de la Plata to Cape Horn, and from Cape Horn to Valparaiso, including the Isles of Juan Fernandes*, 1802. Repository: John Carter Brown Map Collection.
40. Secret communication from Viceroy Croix to Viceroy Loreto, Lima 16 April 1788 and secret answer dated in Buenos Aires 16th June, same year, General Archive of the Indies, Buenos Aires Audience 99.
41. G. Hassel, *Statistisher Umriss der sämmtlichen Europäischen und der vornehmsten aussereuropäischen Staaten* […], Vol. 3, 1824, Repository: Florence Central National Library. Google Books. *Diccionario Geográfico Universal, T.4*, Madrid, 1831; C. Malte-Brun, *Précis de la géographie universelle: ou Description de toutes, les parties du monde sur un plan* […], Vol. 11, 1835. Repository: University of Michigan—Digital.
42. Ch. Reichard, *Süd America*, Weimar, 1804. Repository: National Library Mariano Moreno, Argentine Republic.
43. J. A. C Buchon and H. C. Carey, Carte geographique, statistique et historique de l'Amerique Meridionale, 1825; C. F. Weiland, *Geographisch-statistiche und historische Charte von Patagonien, Feuerland and die Südpolarländer*, 1828; W. Schlieben, *Britische Reich*, 1830. Repository: David Rumsey Map Collection.
44. C. F. Weiland, *Geographisch-statistiche und historische Charte von Patagonien, Feuerland and die Südpolarländer*, 1828. Repository: David Rumsey Map Collection.
45. The instructions given to Fitz Roy in 1831 insisted on the fact that "though disappointing and barren the coast may be, discovering a port that provides shelter and healthy provisions can turn it into the most precious". He was also warned about "our current ignorance of the Falkland Islands, although often visited" and the necessity of accompanying charts and plans with accurate instructions to access the most favorable ports (Fitz Roy 2013).
46. Parker King cites the accounts of García de Loaysa (1525), Sarmiento de Gamboa (1579), Narborough (1652), Byron (1765), Wallis and Carteret (1767), Bougainville (1767) and Córdoba (1785).
47. R. Fitz Roy, *Narrative of the surveying voyages of His Majesty's Ships Adventure and Beagle between the years 1826 and 1836 describing their examination of the Southern Shores of South America, and the Beagle's circumnavigation of the Globe, in three volumes*, London, Henry Colburn, Great Marlborough Street, 1839. It includes the accounts of the first and second expedition and Charles Darwin's "Journal and Remarks".
48. P. Parker King, *Chart of a Part of South America* […], 1831. Repository Robarts Library, University of Toronto

49. Great Britain - Hydrographic Office, *The Strait of Magalhaens commonly called Magellan* […] 1834. Repository: David Rumsey Map Collection, David Rumsey Map Center, Stanford Libraries.
50. J. Arrowsmith, *The Falkland Island/ surveyed by Captn. Robert Fitz Roy R.N. and the officers of H.M. Ship Beagle,* 1841. Repository: National Library of Australia.
51. British Admiralty, *The South-Eastern of Tierra del Fuego with Staaten Island, Cap Horn and Diego Ramírez Island* […], 1836. Repository: Raremaps. Hidrographic Office, 1877, 1st publ. 1841 Royal Museum Greenwich.
52. L. Vernet "*Plano geográfico de la Isla Oriental de Malvinas levantado por D. Luis Vernet sobre los reconocimientos que practicó en los años de 1826, 27 y 28 reducido a punto pequeño de ocho millas por cada pulgada*", 1829. National General Archive (Buenos Aires). Fondo Luis Vernet. Repository: Patagonian Institute of Social and Human Sciences, CCT CONICET CENPAT.
53. Decree creating the Political and Military Command in Malvinas, Buenos Aires, 10th June 1829, signed by Martín Rodríguez and Salvador M. del Carril. National General Archive (Buenos Aires)—Malvinas Islands - *Fondo Luis Vernet, Sala* VII 2-3-3.
54. L. Vernet, *Informe del Comandante Político y Militar de Malvinas,* Buenos Aires, 10 h August 1832. Repository: National Library of Teachers. Argentina Ministry of Education.
55. Manuel Moreno's note to the Minister of Foreign Affairs dated in London 2nd October 1833. National General Archive, Sala X 1-2-7
56. J. Arrowsmith, *The Falkland Island/ surveyed by Captn. Robert Fitz Roy R.N. and the officers of H.M. Ship Beagle, 1834,* 1841. Repository: National Library of Australia.
57. SDUK, *Falkland Isles or Iles Malouines,* 1838. Repository: David Rumsey Map Collection. David Rumsey Map Center, Stanford Libraries.
58. During this period, several expeditions took place successively: W. Smith (1819), F. G. von Bellingshausen (1819–1821), E. Bransfield (1820), J. Weddell (1822–1824), N. Palmer (1820–1821), J. S. C. Dumont d'Urville (1837–1840), Ch. Wilkes (1838–1842) and J.C. Ross (1839–1843).
59. H. Hondius, "*Poli Arctici, et Circumiacentum Terrarum Descriptio Novissima*", 1600; G. Mercator "*Polus arcticus cum vicinis regionibus*", 1616; Scherer "*Regionum Circum Polarium Lapponiae Islandiae Groenlandiae Novae et Veretis Nova Descriptio Geographica*", 1701.
60. S. J. Neele and J. Wilkes, *Countries surrounding the South Pole,* London, 1807. Repository: National Library of Australia.
61. N. Desmarest, *Carte des glaces circumpolaires australes,* 1827. Repository: David Rumsey Map Collection
62. SDUK, *Circumjacent the South Pole,* 1831. Repository: David Rumsey Map Collection.
63. J. Arrowsmith, *The World on Mercator projection,* 1832. Repository: David Rumsey Map Collection.

64. J. Pinkerton, *Southern Hemisphere*, 1812. Repository: David Rumsey Map Collection
65. J. Thomson, *Southern Hemisphere*, 1814. Repository: David Rumsey Map Collection.
66. SUDK, *Patagonia, S. Shetlands, S. Orkneys*, 1838. Repository: David Rumsey Map Collection.
67. J. Wyld & J. Wyld, *Southern hemisphere*, 1849. Repository: National Library of Australia.
68. J. D. Hooker, *The World of the south circumpolar regions. Showing the Discovery Tracks of Cook [1773-1775], Weddell [1822-1824] and Ross [1939-1843]*, 1844. Repository: Linda Hall Library, Digital Collections.
69. H. Smith Evans, *Emigration map of the Word - Geographical and physical Map of the Word on Mercator projection showing the British possessions […]*, 1849. Repository: National Library of Australia.
70. C. Gay, *Atlas de la Historia física y política*, 1854. Repository: David Rumsey Map Collection.
71. Seelstrang, A. and Tourmente, A., *Mapa de la República Argentina... construido por orden del Comité Central Argentino para la Exposición de Filadelfia,* 1875. Repository: National Library in France, Charts and Maps Department, Digital.
72. R. Napp, *La República Argentina*, Buenos Aires, 1876. Repository: University of Michigan—Digital.
73. Musters was assigned in 1861 to a Royal Navy ship that served at the south-Atlantic coast of America and in 1866 left it to devote to sheep-breeding on lands he bought near Montevideo. Following his voyage across Patagonia, in June 1871, he retired with the rank of *commander* (Musters 1979, R. Rey Balmaceda's preliminary study, pp. 20-21).
74. F. A. Garnier, *Patagonie et Détroit de Magellan, Terres Australes*, 1860. Repository: David Rumsey Map Collection
75. Lois (2018) mentions three Tounens' manuscript maps of the *Reino de la Araucanía y la Patagonia* and reproduces one, published in 1886 by cartographer A. Andri and printer C. Frédicx, which depicts it from the Bio Bio and Colorado Rivers up to the southern end of the continent.
76. The map *South America* (1861) of the English cartographer K. Johnson is one of the few that represents Chilean jurisdiction over all Patagonia to the south of the Negro River, (Lacoste 2003: 240), although other maps suggest different ideas and express Argentinean claims.
77. P. E. Coni, *Mapa de la República Argentina y de las Repúblicas Oriental del Uruguay, Paraguay y Chile*, 1868 y 1874. Repository: The Barry Lawrence Ruderman Collection, David Rumsey Map Center, Stanford Libraries.
78. The map at the left bottom corner presents similitudes with Cary (1807), Delarochete (1811) and Tanner's marginal map of Patagonia (1836), although these include the Malvinas/Falkland Islands.
79. *Artículo 2. Sin perjuicio de las cesiones que puedan hacerse en congreso general, se declara: que su territorio norte-sud, desde el arroyo del Medio*

hasta la entrada de la Cordillera en el mar, lindando por una línea al oeste-sudoeste, y por el oeste con las faldas de las cordilleras, y por el nordeste y este con los ríos Paraná y Plata, y con el Atlántico, comprendiendo las islas de Martín García y las adyacencias a sus costas fluviales y marítimas (cited in Aramburo 2016, the underlining is ours).

80. Following Juan Manuel de Rosas fall at Caseros battle (1852), Buenos Aires declined to participate subordinately in the organization project led by Justo José de Urquiza from Paraná and rejected its loss of control of its customs and port. Thus, after rejecting the National Constitution (1853), although not openly refusing to being part of the Argentine Nation, in 1854 Buenos Aires waived its political status as a province and substituted it by the status of State, declaring its autonomy and its exclusive management of foreign relations, defining its territorial dominion as well (Aramburo 2016).
81. In 1876, on the basis of this bill, Law 817 on Immigration and Settlement, known as Avellaneda Law, was promulgated.
82. Note from Minister Felipe Arana to the Ministry of Foreign Relations in Chile dated on 15 February 1847, cited by Pelliza (1969).
83. We refer to the expeditions of F. P. Moreno (1876-1877), R. Lista (1877- 1878), J. T. Rogers (1877–1879), R. Serrano Montaner (1879) and C. M. Moyano (1879–1880).
84. C. Gay, *Estrecho de Magallanes*, in: *Atlas de la Historia física y política*, 1854. Repository: Peter H. Raven Library, Missouri Botanical Garden—Digital.
85. D. Barros Arana and M. J. Olavarrieta, *Chile construido en vista de las mejores cartas i exploraciones oficiales*, Santiago, 1872. Repository: National Library Chile Republic– Digital.
86. R. Napp *La República Argentina*, Buenos Aires, 1876. Repository: University of Michigan. This work rectifies the demarcation of the Magellan region on the map of A. Seelstrang and A. Tourmente (1875) that fails to include Brunswick Peninsula, where the Chilean colony of Punta Arenas was settled, accounting for the error on the fact that "they only had forty days to have the map ready and that prevented them from appropriately looking for information in the archives".
87. *Missioni Salesiane della Patagonia Settentrionale* y *Missioni Salesiane della Patagonie centrale e Meridionle e della Terra del Fuoco,* 1888. Repository: Universiteits bibliotheek Gent.
88. M. Prieto, *Carta que demuestra las diversas proposiciones de arreglo de la cuestión de límites chileno-argentina*, Santiago, 1881. Repository: National Library Mariano Moreno. Argentine Republic.
89. C. Soza Bruna, *Carta de la Extremidad Meridional de Sud América construida para señalar el límite entre Chile y la República Argentina*, 1889. Reproduced in Martinic (1999).
90. Chile organized the Magellan Colonization Territory during 1850 and 1860 decades and Argentina created the National Territory of Patagonia in 1878. Later, by Act 1532 of 1884, it divided the frontier territories in nine governorates: in the north, Chaco, Formosa and Misiones, and in the south, Chubut,

Neuquén, Pampa, Río Negro, Santa Cruz and Tierra del Fuego and the Malvinas Islands.

91. M. F. Paz Soldán, *Gobernación de la Tierra del Fuego y las Islas Malvinas*, Argentine Geographical Atlas, Buenos Aires, 1888. Repository: David Rumsey Map Collection.
92. The Letters Patent of 1908 formalized British jurisdiction by declaring that South Georgia, South Orkneys, South Shetlands and Sandwich archipelagos and a sector of the Antarctic Peninsula, Graham Land, as Falkland Dependences.
93. Regarding the direct link with the international market and the little dependence on the respective national decision-making centers—Buenos Aires and Santiago—Barbería (1991) and Martinic (1992) introduced the concept of "autarchic region". Perhaps the idea of "corporate territory", as developed by Silveira (2007), might be more convenient to think of the privatized national territories functioning.
94. G. W. Colton, *Argentine Republic, Chile, Uruguay, Paraguay, Patagonia,* etc.- *Southern Portion of the Peninsula of Patagonia*, 1879. Repository: David Rumsey Map Collection.
95. Admiralty, *The south-eastern part of Tierra del Fuego with Staaten Island, Cape Horn and Diego Ramírez Islands. Surveyed by Captain Robert Fitzroy RN and the officers of HMS Beagle. 1830-1834*, 1841/1877. Repository: Royal Museum Greenwich - Digital.
96. W. Parker Snow, *Chart of the coast of Patagonia The Falkland Islands & Tierra del Fuego, in A two years' cruise off Tierra del Fuego, the Falkland islands, Patagonia, and in the River Plate; a narrative of life in the Southern Seas*, London, 1857. Repository: Robarts Library, University of Toronto—Digital.
97. L. F. Martial, *Carte ethnographique de l'archipel de la Terre de Feu,* 1894. En Mission Scientifique du Cap Horn, 1882-1883, Ministère de la Marine et des Colonies—Ministère de l'Education Nationale, 1885-1891. Repository: Digitizing sponsor Google - Archive.org.
98. G. Bove, *Terra del Fuoco. Distribuzione delle razze fuegine*, 1883. In *Expedición Austral Argentina. Informes preliminares presentados a S.S.E.E. los Ministros del Interior y de Guerra y Marina de la República Argentina.* Repository: Smithsonian Libraries—Digital.
99. A. Torrealba, *Mapa de la región austral Chile. Provincias de Llanquihue, Chiloé y Territorio de Magallanes*, Boundaries Demarcation and Colonization and Lands General Inspection Office, Santiago, 1904. Repository: Magellan University—Aike Digital Library of Patagonia. Reproduced by Martinic 1999: 155.
100. Lands and Colonies Department, *Plano demostrativo del estado de la tierra pública en los Territorios Nacionales del Sud*, Buenos Aires, 1900. Repository: National Library Mariano Moreno. Argentine Republic.

References

Aramburo M (2016) El debate legislativo de la constitución del Estado de Buenos Aires (1854). Los conceptos de soberanía, nación y Estado. PolHis 9(17):170–209

Bandieri S (2005) *Historia de la Patagonia.* Buenos Aires, Sudamericana

Barbería EM (1991) Chile y Argentina: una región autárquica en el sur. 1880-1920. Waxen, Publicación Científica Universidad Federal de la Patagonia Austral, V 4:3–43

Basberg BL (2017) Commercial and economic aspects of Antarctic exploration—from the earliest discoveries into the Heroic Age. The Polar Journal 7(1):205–226

Bascopé Julio J (2008) Pasajeros del poder propietario. La Sociedad Explotadora de Tierra del Fuego y la biopolítica estanciera (1890-1920). Magallania (Chile) 36(2):19–44

Bascopé Julio J (2010) Sentidos coloniales I. El oro y la vida salvaje en Tierra del Fuego, 1880-1914. Magallania (Chile) 38(2):5–26

Bascopé Julio (2011) "Bajo tuición. Infancia y extinción en la historia de la colonización fueguina. (Sentidos coloniales II)", *Corpus* 1(1)

Bascopé Julio (2016) *El encantador de yaganes. Entrenamiento de nativos fueguinos en la isla Keppel, 1854–1869*, Buenos Aires, CLACSO

Belfiori MC (1977) Tierra del Fuego. Destino de la tierra pública. Karukinka Cuaderno Fueguino 19–20:4–15

Bennassar B (1987) *La América española y la América portuguesa.* Siglos XVI-XVIII, Madrid, Akal

Berguño F (2002) Proyectos franceses de colonización de las islas Malvinas, el Estrecho de Magallanes y la Patagonia (1821-1843). Anales Instituto Patagonia 30:17–26

Boccara G, Seguel-Boccara I (1999) Políticas indígenas en Chile (siglos XIX y XX). De la asimilación al pluralismo (El caso mapuche). Revista de Indias, LIX 217:741–774

Boccara, G. (2002) "Colonización, resistencia y etnogénesis en las fronteras americanas", Boccara, G. (2002) (ed.), *Colonización, resistencia y mestizaje en las Américas (Siglos XVI-XX)*, Quito, Abya-Yala, pp. 47–82

Boccara, G. (2003) "Fronteras, mestizaje y etnogénesis", Mandrini, R. J. and Paz, C. D. (comp.), *Las fronteras hispano-criollas del mundo indígena latinoamericano en los siglos XVIII y XIX. Un estudio comparativo*, Tandil, Universidad Nacional del Centro de la Provincia de Buenos Aires—Universidad Nacional del Comahue, pp. 63–108

Bridges EL (1978) *El último confín de la Tierra.* Buenos Aires, Marymar

Bulkeley R (2016) Naming Antarctica. Polar Record 52(262):2–15

Caillet-Bois R (1982) *Una tierra argentina.* Buenos Aires, Academia Nacional de la Historia

Cain MT (1994) The Maps of the Society for the Diffusion of Useful Knowledge: A Publishing History. Imago Mundi 46:151–167

Coronato F (2016) *Ovejas y ovejeros en la Patagonia.* Buenos Aires, Prometeo

D'Orbigny A (1999) *Viaje por América meridional II*. Buenos Aires, Emecé

Ferns HS (1984) *Gran Bretaña y Argentina en el siglo XIX*. Buenos Aires, Solar

Fitz Roy R (2013) *Viajes del Adventure y el Beagle.* Diario, Madrid, CSIC

Garavaglia, J. C. (2007) *Construir el estado, inventar la nación. El Río de la Plata, siglos XVIII-XIX*, Buenos Aires, Prometeo

García SV (2010) La logística de los levantamientos hidrográficos en el Río de la Plata y Patagonia en tiempos del HMS Beagle. Anuario IEHS 25:289–301

González Bollo H (1998) Una tradición de cartografía física y política de la Argentina, 1838-1882. Ciencia Hoy 8(46):12–23

González Bollo H (1999) Estado, ciencia y sociedad: los manuales estadísticos y geográficos en los orígenes de la Argentina moderna, 1852-1876. Anuario del IEHS 14:23–42

Harambour RA (2015) El ovejero y el bandido. Trayectorias, cruces y genocidio en dos relatos de viaje británicos en Tierra del Fuego (década de 1890). Anales de Literatura Chilena 16(24):163–182

Harambour RA (2016) "Sheep Sovereignties: The Colonization of the Falkland Island/ Malvinas, Patagonia, and Tierra del Fuego, 1830 s-1910 s", Beezley, W. (ed.), *Oxford Research Encyclopedia of Latin American History*, Nueva York, Oxford University Press

Harambour RA (2017) "Soberanía y corrupción. La construcción del Estado y la propiedad en Patagonia austral (Argentina y Chile, 1840-1920). Historia, II 50:555–596

Harambour RA, Barrena Ruiz J (2019) Barbarie o justicia en la Patagonia occidental: las violencias coloniales en el caso del pueblo kawésqar, finales del siglo XIX e inicios del siglo XX. Historia Crítica 71:25–48

Harley JB (2005) *La nueva naturaleza de los mapas.* Ensayos sobre la historia de la cartografía, México, Fondo de Cultura Económica

Hobbs WH (1940) The Discovery of Wilkes Land, Antarctica. Proceedings of American Philosophical Society 82(5):561–582

Jones AGE (1981) The British Southern Whale and Seal Fisheries. Part II: The Principal Operators. The Great Circle 3(2):90–102

Lacoste P (2003) *La imagen del otro en las relaciones de la Argentina y Chile (1534-2000).* Buenos Aires, Fondo de Cultura Económica

Levaggi A (1978) "El presidio español de las islas Malvinas", *Investigaciones y Ensayos*, 24. Buenos Aires, Academia Nacional de la Historia

Levaggi (2000) *Paz en la frontera. Historia de las relaciones diplomáticas con las comunidades indígenas en la Argentina (Siglos XVI—XIX)*, Buenos Aires, Dunken

Lois, C. (2007) "La Patagonia en el mapa de la Argentina moderna. Política y "deseo territorial" en la cartografía oficial argentina en la segunda mitad del siglo XIX", Navarro Floria, P. (coord.), *Paisajes del progreso: la resignificación de la Patagonia Norte, 1880–1916*, Neuquén, EDUCO—Universidad Nacional del Comahue

Lois C (2010) "Las evidencias, lo evidente y lo visible: el uso de dispositivos visuales en la argumentación diplomática argentina sobre la Cordillera de los Andes (1900) como frontera natural", *Treballs de la Societat Catalana de Geografia*, 70

Lois C (2015) "Trazos a ciegas: los mapas políticos de Sudamérica en tiempos de las revoluciones independentistas latinoamericanas", Roque Oliveira, F. (ed.), *Cartógrafos para toda la Tierra. Producción y circulación del saber cartográfico iberoamericano: agentes y contextos*, Lisboa, Vol. 1, pp 287–316

C Lois (2018) *Terrae incognitae.* Modos de pensar y mapear geografías desconocidas, Buenos Aires, Eudeba

Luiz, M. T. (2006) *Relaciones fronterizas en Patagonia: la convivencia hispano-indígena a fines del período colonial*, Ushuaia, Asociación HANIS—Universidad Nacional de la Patagonia San Juan Bosco

Llanos Reyes C (2010) Imperialismo inglés y ciencia. La Sociedad Geográfica Real de Londres, 1830-1870. Boletín Americanista 60:209–225

Manara, C. G. (2005) "La frontera surandina: centro de la confrontación política a principios del siglo XIX", *Mundo Agrario. Revista de estudios rurales*, 5 (10)

Mandrini R (1992) Indios y fronteras en el área pampeana (siglos XVI- XIX). Balance y perspectivas. Anuario del IEHS 7:59–73

Martinic BM (1979) La política indígena de los gobernadores de Magallanes 1843-1910. Anales del Instituto de la Patagonia 10:7–58

Martinic BM (1980) La misión de Bayly (archipiélago del Cabo de Hornos). Anales del Instituto de la Patagonia 11:47–61

Martinic BM (1992) *Historia de la Región Magallánica.* Universidad de Magallanes, Punta Arenas

Martinic BM (1999) *Cartografía Magallánica, 1523-1945.* Universidad de Magallanes, Punta Arenas

Martinic (2001) Patagonia austral: 1885–1925. Un caso singular y temprano de integración regional autárquica. Bandieri, S. (coord.), *Cruzando la Cordillera. La frontera argentino-chilena como espacio social*, Neuquén, CEHIR—Universidad Nacional del Comahue, pp 459–486

Martinic (2009) Relaciones y comercio entre Magallanes y las islas Falkland (1845–1950), *Magallania*, 37(2): 5–21
Miller RJ, LeSage L, López Escarcena S (2010) The International Law of Discovery, Indigenous Peoples, and Chile. Nebraska Law Review 89(4):819–884
Moreno Alonso M (1983) Las islas del Atlántico Sur y el imperialismo británico en el siglo XIX. Anuario de Estudios Americanos, XL, pp 313–357
Musters, G. Ch. (1979) *Vida entre los patagones. Un año de excursiones por tierras no frecuentadas, desde el Estrecho de Magallanes hasta el río Negro*, Buenos Aires, Solar/Hachette
Navarro Floria P (2002) El desierto y la cuestión del territorio en el discurso político argentino sobre la frontera sur. Revista Complutense de Historia de América 28:139–168
Navarro Floria P (2003) Córdoba y Malaspina: antropología y política ilustrada en Patagonia y Tierra del Fuego. Revista Española de Antropología Americana 33:231–251
Navarro Floria P (2005) La Patagonia en la clasificación del hombre: el desencantamiento de los "patagones" y su aporte a la historia de la Antropología. Revista Española de Antropología Americana 35:169–189
Nicoletti MA (2008) *Indígenas y misioneros en la Patagonia: huellas de los salesianos en la cultura y la religiosidad de los pueblos originarios*. Buenos Aires, Continente
Otero H. (1997–1998) "Estadística censal y construcción de la nación. El caso argentino, 1869–1914", *Boletín del Instituto Ravignani*, 16–17: 123-149
Pagden A (1988) *La caída del hombre natural*. El indio americano y los orígenes de la etnología comparativa, Madrid, Alianza
Parish W (1833) Account of East Falkland Island. Journal of the Royal Geographical Society, London 3:94–99
Parish W (1958) *Buenos Aires y las Provincias del Río de la Plata*. Buenos Aires, Solar/Hachette
Parker King P (2015) *Los viajes del Beagle. Informes de la Primera Expedición (1826–1830)*, Buenos Aires, Eudeba
Passetti G (2014) Strait of Magellan rediscovered: science, politics, commerce and the exploring expeditions during the 1820's and 1830's. Antíteses 7(13):254–276
Pelliza MA (1969) *La cuestión del Estrecho de Magallanes*. Buenos Aires, Eudeba
Quijada M (1999) La ciudadanización del "indio bárbaro". Políticas oficiales y oficiosas hacia la población indígena de la pampa y la Patagonia, 1870-1920. Revista de Indias, LIX 217:675–704
Quijada M (2000) Nación y territorio: la dimensión simbólica del espacio en la construcción nacional argentina, siglo XIX. Revista de Indias, LX 219:373–394
Quijada M (2002) Pensando la frontera sur argentina: concepto, contenido, continuidades y discontinuidades de una realidad espacial y étnica (Siglos XVIII-XIX). Revista de Indias, LXII 224:103–142
Quintero Palacios S (1995) Geografía y Nación. Estrategias Educativas en la representación del territorio argentino. *Territorio* 7, Publicaciones del Instituto de Geografía, Universidad de Buenos Aires
Ratto S (2003) Cuando las fronteras se diluyen. Las formas de interrelación blanco-indias en el sur bonaerense. In: Mandrini, R. J. and Paz, C. D. (comp.), *Las fronteras hispano-criollas del mundo indígena latinoamericano en los siglos XVIII y XIX. Un estudio comparativo*, Tandil, Universidad Nacional del Centro de la Provincia de Buenos Aires—Universidad Nacional del Comahue, pp. 199–232
Rose JH (1929) Captain Cook and the Founding of British Power in the Pacific. Geogr J 73(2):102–122
Sagredo Baeza R (2018) *El futuro de Chile delineado en un mapa*. Revista de Geografía Norte Grande 69:49–69
Senatore MX, Zarankin A (1999) Arqueología histórica y expansión capitalista. Prácticas cotidianas y grupos operarios en la Península Byres, Isla Livingston, Shetland del Sur. Zarankin, A. and Acuto F. A. (eds.), *Sed Non Satiata. Teoría Social en la Arqueología Latinoamericana*, Buenos Aires, Ediciones del Tridente, pp 171–188
Serrano A (2006) *Memorias recientes de la región del Cabo de Hornos*. Punta Arenas, FONDART

Silveira ML (2007) Los territorios corporativos de la globalización. Geograficando, Revista de Estudios Geográficos 3(3):13–26

Silvestri G (2011) *El lugar común. Una historia de las figuras de paisaje en el Río de la Plata*, Buenos Aires, Edhasa

Staszak JF (2012) La construcción del imaginario occidental del "allá" y la fabricación de las "exótica": el caso de los *toi moko* maoríes. In: Lindón, A. and Hiernaux. D. (dirs.), *Geografías de lo imaginario*, Barcelona, Anthropos—México, Universidad Autónoma Metropolitana, pp 177–203

Warnick S (2008) *The reluctant colonization of the Falkland Islands, 1833–1851: a study of British Imperialism in the Southwest Atlantic*, Unpublished Master's Theses, Paper 702, University of Richmond

Weber DJ (1998) Borbones y Bárbaros. Centro y periferia en la reformulación de la política de España hacia los indígenas no sometidos. *Anuario del IEHS*, 13: 147–171, Universidad Nacional del Centro de la Provincia de Buenos Aires

Weddell J (2006) *Un viaje hacia el Polo Sur realizado en los años 1822-1824.* Buenos Aires, Eudeba

Yan S (2007) Mapping Knowledge and Power: Cartographic Representations of Empire in Victorian Britain. EurAmerica 37(1):1–34

Yuln M (2014) Representaciones gráficas del Estado en formación. La tierra y los indígenas en la cartografía territorial argentina durante el siglo XIX. Almanack. Guarulhos 8:142–155

Zarankin A, Senatore MX (2005) Archaeology in Antarctica: nineteenth-century capitalism expansion strategies. International Journal of Historical Archaeology 9(1):121–150